Learning Ansible 2.7
Third Edition

Automate your organization's infrastructure using Ansible 2.7

Fabio Alessandro Locati

BIRMINGHAM - MUMBAI

Learning Ansible 2.7
Third Edition

Commissioning Editor: Karan Sadawana
Acquisition Editor: Prachi Bisht
Content Development Editor: Drashti Panchal
Technical Editor: Komal Karne
Copy Editor: Safis Editing
Project Coordinator: Jagdish Prabhu
Proofreader: Safis Editing
Indexer: Manju Arasan
Graphics: Tom Scaria
Production Coordinator: Nilesh Mohite

First published: November 2014
Second edition: November 2016
Third edition: April 2019

Production reference: 1300419

Published by Packt Publishing Ltd.
Livery Place
35 Livery Street
Birmingham
B3 2PB, UK.

ISBN 978-1-78995-433-3

www.packtpub.com

`mapt.io`

Mapt is an online digital library that gives you full access to over 5,000 books and videos, as well as industry leading tools to help you plan your personal development and advance your career. For more information, please visit our website.

Why subscribe?

- Spend less time learning and more time coding with practical eBooks and Videos from over 4,000 industry professionals

- Improve your learning with Skill Plans built especially for you

- Get a free eBook or video every month

- Mapt is fully searchable

- Copy and paste, print, and bookmark content

Packt.com

Did you know that Packt offers eBook versions of every book published, with PDF and ePub files available? You can upgrade to the eBook version at `www.packt.com` and as a print book customer, you are entitled to a discount on the eBook copy. Get in touch with us at `customercare@packtpub.com` for more details.

At `www.packt.com`, you can also read a collection of free technical articles, sign up for a range of free newsletters, and receive exclusive discounts and offers on Packt books and eBooks.

Contributors

About the author

Fabio Alessandro Locati, commonly known as Fale, is a director at Otelia, a public speaker, an author, and an open source contributor. His main areas of expertise are Linux, automation, security, and cloud technologies. Fale has more than 12 years of working experience in IT, with many of them spent consulting for many companies, including dozens of Fortune 500 companies. This has allowed him to consider technologies from different points of view, and to think critically about them.

I would like to thank my parents, who introduced me to computer science before I was even able to write, and my whole family, who has always been supportive. A special thanks goes to everyone I worked with at Packt for their hard work, and to Timothy Rupp for his great feedback. Since Ansible is an open source project, I thank all the companies that decided to invest in it, and all the people that decided to volunteer their time to the project.

About the reviewer

Timothy Rupp has been working in various fields of computing for the last 15 years. He has held positions in cybersecurity, software engineering, and in the cloud computing and DevOps fields.

He was first introduced to Ansible while at Rackspace. As part of the cloud engineering team, he made extensive use of the tool to deploy new capacity for Rackspace Public Cloud. Since that introduction, he has contributed patches, provided support, and presented on Ansible topics at local meetups.

While at F5 Networks, he led the development of F5's Ansible modules, and became a core contributor to the Ansible project. Most recently, he has become re-involved in cybersecurity in the financial sector.

Packt is searching for authors like you

If you're interested in becoming an author for Packt, please visit `authors.packtpub.com` and apply today. We have worked with thousands of developers and tech professionals, just like you, to help them share their insight with the global tech community. You can make a general application, apply for a specific hot topic that we are recruiting an author for, or submit your own idea.

Table of Contents

Preface

The information technology sector is a fast-moving sector that always tries to accelerate. To keep up with this, companies need to be able to move quickly and iterate frequently. Until a few years back, this was mainly true for software, but now we start to see the necessity of developing infrastructure at a similar speed. Going forward, we will need to change the infrastructure we run our software on at the speed of the software itself.

In this scenario, many technologies, such as software-defined everything (storage, network, and compute, for example), will be key, but those technologies need to be managed in an equally scalable way, and that way will involve using Ansible and similar products.

Ansible is highly relevant today, since, unlike competing products, it is agentless, allowing for faster deployments, more security, and better auditability.

Who this book is for

The book is for developers and sysadmins who want to automate their organization's infrastructure using Ansible 2. No prior knowledge of Ansible is required.

What this book covers

Chapter 1, *Getting Started with Ansible*, explains how to install Ansible.

Chapter 2, *Automating Simple Tasks*, explains how to create simple playbooks that will allow you to automate some simple tasks that you already perform on a daily basis.

Chapter 3, *Scaling to Multiple Hosts*, explains how to handle multiple hosts in Ansible in an easy-to-scale way.

Chapter 4, *Handling Complex Deployment*, explains how to create deployments that have multiple phases, as well as multiple machines.

Chapter 5, *Going Cloud*, explains how Ansible can integrate with various cloud offerings, and how it can simplify your life, managing the cloud for you.

Chapter 6, *Getting Notification from Ansible*, explains how to set up Ansible to return valuable information to you and other stakeholders.

Chapter 7, *Creating a Custom Module,* explains how to create a custom module to leverage the freedom Ansible gives you.

Chapter 8, *Debugging and Error Handling,* explains how to debug and test Ansible to ensure that your playbooks will always work.

Chapter 9, *Complex Environments,* explains how to manage multiple tiers, environments, and deployments with Ansible.

Chapter 10, *Introducing Ansible for Enterprises,* explains how to manage Windows nodes from Ansible, as well as how to leverage Ansible Galaxy to maximize your productivity.

Chapter 11, *Getting Started with AWX,* explains what AWX is and how you can start using it.

Chapter 12, *Working with AWX Users, Permissions, and Organizations,* explains how AWX users and permissions management works.

To get the most out of this book

This book assumes a basic knowledge of the UNIX shell, as well as basic networking knowledge.

Download the example code files

You can download the example code files for this book from your account at www.packt.com. If you purchased this book elsewhere, you can visit www.packt.com/support and register to have the files emailed directly to you.

You can download the code files by following these steps:

1. Log in or register at www.packt.com.
2. Select the **SUPPORT** tab.
3. Click on **Code Downloads & Errata**.
4. Enter the name of the book in the **Search** box and follow the onscreen instructions.

Once the file is downloaded, please make sure that you unzip or extract the folder using the latest version of:

- WinRAR/7-Zip for Windows

- Zipeg/iZip/UnRarX for Mac
- 7-Zip/PeaZip for Linux

The code bundle for the book is also hosted on GitHub at `https://github.com/PacktPublishing/Learning-Ansible-2.X-Third-Edition`. In case there's an update to the code, it will be updated on the existing GitHub repository.

We also have other code bundles from our rich catalog of books and videos available at `https://github.com/PacktPublishing/`. Check them out!

Conventions used

There are a number of text conventions used throughout this book.

`CodeInText`: Indicates code words in text, database table names, folder names, filenames, file extensions, pathnames, dummy URLs, user input, and Twitter handles. Here is an example: "The `sudo` command is a well known, but is often used in its more dangerous form."

A block of code is set as follows:

```
- hosts: all
  remote_user: vagrant
  tasks:
    - name: Ensure the HTTPd package is installed
      yum:
        name: httpd
        state: present
      become: True
```

When we wish to draw your attention to a particular part of a code block, the relevant lines or items are set in bold:

```
- hosts: all
  remote_user: vagrant
  tasks:
    - name: Ensure the HTTPd package is installed
      yum:
        name: httpd
        state: present
      become: True
```

Any command-line input or output is written as follows:

```
$ sudo dnf install ansible
```

Warnings or important notes appear like this.

Tips and tricks appear like this.

Get in touch

Feedback from our readers is always welcome.

General feedback: If you have questions about any aspect of this book, mention the book title in the subject of your message and email us at customercare@packtpub.com.

Errata: Although we have taken every care to ensure the accuracy of our content, mistakes do happen. If you have found a mistake in this book, we would be grateful if you would report this to us. Please visit www.packt.com/submit-errata, selecting your book, clicking on the Errata Submission Form link, and entering the details.

Piracy: If you come across any illegal copies of our works in any form on the Internet, we would be grateful if you would provide us with the location address or website name. Please contact us at copyright@packt.com with a link to the material.

If you are interested in becoming an author: If there is a topic that you have expertise in and you are interested in either writing or contributing to a book, please visit authors.packtpub.com.

Reviews

Please leave a review. Once you have read and used this book, why not leave a review on the site that you purchased it from? Potential readers can then see and use your unbiased opinion to make purchase decisions, we at Packt can understand what you think about our products, and our authors can see your feedback on their book. Thank you!

For more information about Packt, please visit packt.com.

Section 1: Creating a Web Server Using Ansible

This section will help you create simple playbooks that will allow you to automate some simple tasks that you already perform on a daily basis.

This section contains the following chapters:

- Chapter 1, *Getting Started with Ansible*
- Chapter 2, *Automating Simple Tasks*

Getting Started with Ansible

1

Information and communications technology (**ICT**) is often described as a fast-growing industry. I think that the best quality of the ICT industry is not related to its ability to grow at a super-high speed, but is related to its ability to revolutionize itself, and the rest of the world, at an astonishing pace.

Every 10 to 15 years there are major shifts in how this industry works, and every shift solves problems that were very hard to manage up to that point, creating new challenges. Also, at every major shift, many of the best practices of the previous iteration are classified as anti-patterns, and new best practices are created. Although it might appear that those changes are impossible to predict, this is not always true. Obviously, it is not possible to know exactly what changes will occur and when they will take place, but looking at companies with a large number of servers and many lines of code usually reveals what the next steps will be.

The current shift has already happened in big companies such as Amazon Web Services (AWS), Facebook, and Google. It is the implementation of IT automation systems to create and manage servers.

In this chapter we will cover the following topics:

- IT automation
- What is Ansible?
- The secure shell
- Installing Ansible
- Creating a test environment with Vagrant
- Version control systems
- Using Ansible with Git

Technical requirements

To support the learning of Ansible, I suggest having a machine where you can install Vagrant. Using Vagrant will allow you to try many operations, even destructive ones, without fear.

Additionally, AWS and Azure accounts are suggested, since some examples will be on those platforms.

All examples in this book are available in the GitHub repository at `https://github.com/PacktPublishing/-Learning-Ansible-2.X-Third-Edition/`.

IT automation

IT automation is – in its broader sense – the processes and software that help with the management of the IT infrastructure (servers, networking, and storage). In the current shift, we are supporting for a huge implementation of such processes and software.

At the beginning of IT history, there were very few servers, and a lot of people were needed to make them work properly, usually more than one person for each machine. Over the years, servers became more reliable and easier to manage, so it was possible to have multiple servers managed by a single system administrator. In that period, the administrators manually installed the software, upgraded the software manually, and changed the configuration files manually. This was obviously a very labor-intensive and error-prone process, so many administrators started to implement scripts and other means to make their lives easier. Those scripts were (usually) pretty complex, and they did not scale very well.

In the early years of this century, data centers started to grow a lot due to companies' needs. Virtualization helped in keeping prices low, and the fact that many of these services were web services meant that many servers were very similar to each other. At this point, new tools were needed to substitute the scripts that were used before: the configuration management tools.

CFEngine was one of the first tools to demonstrate configuration management capabilities way back in the 1990s; more recently, there has been Puppet, Chef, and Salt, besides Ansible.

Advantages of IT automation

People often wonder if IT automation really brings enough advantages, considering that implementing it has some direct and indirect costs. The main benefits of IT automation are the following:

- The ability to provision machines quickly
- The ability to recreate a machine from scratch in minutes
- The ability to track any change performed on the infrastructure

For these reasons, it's possible to reduce the cost of managing the IT infrastructure by reducing the repetitive operations often performed by system administrators.

Disadvantages of IT automation

As with any other technology, IT automation does come with some disadvantages. From my point of view, these are the biggest disadvantages:

- Automating all of the small tasks that were once used to train new system administrators.
- If an error is performed, it will be propagated everywhere.

The consequence of the first is that new ways to train junior system administrators will need to be implemented.

The second one is trickier. There are a lot of ways to limit this kind of damage, but none of those will prevent it completely. The following mitigation options are available:

- **Always have backups**: Backups will not prevent you from nuking your machine – they will only make the restore process possible.
- **Always test your infrastructure code (playbooks/roles) in a non-production environment**: Companies have developed different pipelines to deploy code, and those usually include environments such as dev, test, staging, and production. Use the same pipeline to test your infrastructure code. If a buggy application reaches the production environment it could be a problem. If a buggy playbook reaches the production environment, it can be catastrophic.

- **Always peer-review your infrastructure code**: Some companies have already introduced peer-reviews for the application code, but very few have introduced it for the infrastructure code. As I was saying in the previous point, I think that infrastructure code is way more critical than application code, so you should always peer-review your infrastructure code, whether you do it for your application code or not.

- **Enable SELinux**: SELinux is a security kernel module that is available on all Linux distributions (it is installed by default on Fedora, Red Hat Enterprise Linux, CentOS, Scientific Linux, and Unbreakable Linux). It allows you to limit users and process powers in a very granular way. I suggest using SELinux instead of other similar modules (such as AppArmor) because it is able to handle more situations and permissions. SELinux will prevent a huge amount of damage because, if correctly configured, it will prevent many dangerous commands from being executed.

- **Run the playbooks from a limited account**: Even though user and privilege escalation schemes have been in Unix code for more than 40 years, it seems as if not many companies use them. Using a limited user for all your playbooks, and escalating privileges only for commands that need higher privileges, will help prevent you from nuking a machine while trying to clean an application temporary folder.

- **Use horizontal privilege escalation**: The `sudo` command is a well known, but is often used in its more dangerous form. The `sudo` command supports the `-u` parameter that will allow you to specify a user that you want to impersonate. If you have to change a file that is owned by another user, please do not escalate to `root` to do so, just escalate to that user. In Ansible, you can use the `become_user` parameter to achieve this.

- **When possible, don't run a playbook on all your machines at the same time**: Staged deployments can help you detect a problem before it's too late. There are many problems that are not detectable in dev, test, staging, and QA environments. The majority of them are related to a load that is hard to emulate properly in those non-production environments. A new configuration you have just added to your Apache HTTPd or MySQL servers could be perfectly OK from a syntax point of view, but disastrous for your specific application under your production load. A staged deployment will allow you to test your new configuration on your actual load without risking downtime in case something was wrong.

- **Avoid guessing commands and modifiers**: A lot of system administrators will try to remember the right parameter, and try to guess if they don't remember it exactly. I've done it too, a lot of times, but this is very risky. Checking the man page or the online documentation will usually take you less than two minutes, and often, by reading the manual, you'll find interesting notes you did not know. Guessing modifiers is dangerous because you could be fooled by a non-standard modifier (that is, `-v` is not a verbose mode for `grep`, and `-h` is not a `help` command for the MySQL CLI).

- **Avoid error-prone commands**: Not all commands have been created equally. Some commands are (way) more dangerous than others. If you can assume a `cat`-based command safe, you have to assume that a `dd`-based command is dangerous, since it performs copies and conversion of files and volumes. I've seen people using `dd` in scripts to transform DOS files to Unix (instead of `dos2unix`) and many other, very dangerous, examples. Please, avoid such commands, because they could result in a huge disaster if something goes wrong.

- **Avoid unnecessary modifiers**: If you need to delete a simple file, use `rm ${file}`, not `rm -rf ${file}`. The latter is often performed by users that have learned *to be sure, always use* `rm -rf`, because at some time in their past, they have had to delete a folder. This will prevent you from deleting an entire folder if the `${file}` variable is set wrongly.

- **Always check what could happen if a variable is not set**: If you want to delete the contents of a folder and you use the `rm -rf ${folder}/*` command, you are looking for trouble. In case the `${folder}` variable is not set for some reason, the shell will read a `rm -rf /*` command, which is deadly (considering the fact that the `rm -rf /` command will not work on the majority of current OSes because it requires a `--no-preserve-root` option, while the `rm -rf /*` will work as expected). I'm using this specific command as an example because I have seen such situations: the variable was pulled from a database which, due to some maintenance work, was down, and an empty string was assigned to that variable. What happened next is probably easy to guess. In case you cannot prevent using variables in dangerous places, at least check them to see whether they are empty or not before using them. This will not save you from every problem, but may catch some of the most common ones.

- **Double-check your redirections**: Redirections (along with pipes) are the most powerful elements of Unix shells. They could also be very dangerous: a `cat /dev/rand > /dev/sda` can destroy a disk even if a `cat`-based command is usually overlooked because it's not usually dangerous. Always double-check all commands that include a redirection.

- **Use specific modules wherever possible**: In this list, I've used shell commands because many people will try to use Ansible as if it's just a way to distribute them: it's not. Ansible provides a lot of modules and we'll see them in this book. They will help you create more readable, portable, and safe playbooks.

Types of IT automation

There are a lot of ways to classify IT automation systems, but by far the most important is related to how the configurations are propagated. Based on this, we can distinguish between agent-based systems and agent-less systems.

Agent-based systems

Agent-based systems have two different components: a **server**, and a client called **agent**.

There is only one server, and it contains all of the configuration for your whole environment, while the agents are as many as the machines in the environment.

 In some cases, more than one server could be present to ensure high availability, but treat it as if it's a single server, since they will all be configured in the same way.

Periodically, the client will contact the server to see if a new configuration for its machine is present. If a new configuration is present, the client will download it and apply it.

Agent-less systems

In agent-less systems, no specific agent is present. Agent-less systems do not always respect the server/client paradigm, since it's possible to have multiple servers and even the same number of servers and clients. Communications are initialized by the server, which will contact the client(s) using standard protocols (usually via SSH and PowerShell).

Agent-based versus agent-less systems

Aside from the differences previously outlined, there are other contrasting factors that arise because of those differences.

From a security standpoint, an agent-based system can be less secure. Since all machines have to be able to initiate a connection to the server machine, this machine could be attacked more easily than in an agent-less case, where the machine is usually behind a firewall that will not accept any incoming connections.

From a performance point of view, agent-based systems run the risk of having the server saturated, therefore the roll-out could be slower. It also needs to be considered that, in a pure agent-based system, it is not possible to force-push an update immediately to a set of machines. It will have to wait until those machines check-in. For this reason, multiple agent-based systems have implemented out-of-bands ways to implement such features. Tools such as Chef and Puppet are agent-based, but can also run without a centralized server to scale a large number of machines, commonly called **Serverless Chef** and **Masterless Puppet** respectively.

An agent-less system is easier to integrate in an infrastructure that is already present (brownfield) since it will be seen by the clients as a normal SSH connection, therefore no additional configuration is needed.

What is Ansible?

Ansible is an agent-less IT automation tool developed in 2012 by Michael DeHaan, a former Red Hat associate. The Ansible design goals are for it to be minimal, consistent, secure, highly reliable, and easy to learn. The Ansible company was bought by Red Hat in October 2015, and now operates as part of Red Hat, Inc.

Ansible primarily runs in push mode using SSH, but you can also run Ansible using `ansible-pull`, where you can install Ansible on each agent, download the playbooks locally, and run them on individual machines. If there are a large number of machines (large is a relative term; but in this case, consider it to mean greater than 500), and you plan to deploy updates to the machines in parallel, this might be the right way to go about it. As we discussed before, either agent-full and agent-less systems have their pros and cons.

In the next section, we will discuss Secure Shell (SSH), which is a core part of Ansible and the Ansible philosophy.

Secure Shell

Secure Shell (also known as **SSH**) is a network service that allows you to log in and access a shell remotely over a fully encrypted connection. The SSH daemon is today the standard for UNIX system administration, after having replaced the unencrypted telnet. The most frequently used implementation of the SSH protocol is OpenSSH.

In the last few years, Microsoft has implemented OpenSSH for Windows. I think that this proves the *de facto* standard situation that SSH lives into.

Since Ansible performs SSH connections and commands in the same way any other SSH client would do, no specific configuration has been applied to the OpenSSH server.

To speed up default SSH connections, you can always enable `ControlPersist` and the pipeline mode, which makes Ansible faster and more secure.

Why Ansible?

We will try and compare Ansible with Puppet and Chef during the course of this book, since many people have good experience those tools. We will also point out specifically how Ansible would solve a problem compared to Chef or Puppet.

Ansible, as well as Puppet and Chef, are declarative in nature, and are expected to move a machine to the desired state specified in the configuration. For example, in each of these tools, in order to start a service at a point in time and start it automatically on restart, you would need to write a declarative block or module; every time the tool runs on the machine, it will aspire to obtain the state defined in your **playbook** (Ansible), **cookbook** (Chef), or **manifest** (Puppet).

The difference in the toolset is minimal at a simple level, but as more situations arise and the complexity increases, you will start finding differences between the different toolsets. In Puppet, you do not set the order in which the tasks will be executed, and the Puppet server will decide the sequence and the parallelizations at runtime, making it easier to end up with difficult-to-debug errors. To exploit the power of Chef, you will need a good Ruby team. Your team needs to be good at the Ruby language to customize both Puppet and Chef, and there will be a bigger learning curve with both of the tools.

With Ansible, the case is different. It uses the simplicity of Chef when it comes to the order of execution – the top-to-bottom approach – and allows you to define the end state in YAML format, which makes the code extremely readable and easy for everyone, from development teams to operations teams, to pick up and make changes. In many cases, even without Ansible, operations teams are given playbook manuals to execute instructions from whenever they face issues. Ansible mimics that behavior. Do not be surprised if you end up having your project manager change the code in Ansible and check it into Git because of its simplicity!

Installing Ansible

Installing Ansible is rather quick and simple. You can use the source code directly, by cloning it from the GitHub project (`https://github.com/ansible/ansible`); install it using your system's package manager; or use Python's package management tool (**pip**). You can use Ansible on any Windows or Unix-like system, such as macOS and Linux. Ansible doesn't require any databases, and doesn't need to have any daemons running. This makes it easier to maintain Ansible versions and upgrade without any breaks.

We'd like to call the machine where we will install Ansible our Ansible workstation. Some people also refer to it as the command center.

Installing Ansible using the system's package manager

It is possible to install Ansible using the system's package manager, and, in my opinion, this is the preferred option if your system's package manager ships at least Ansible 2.0. We will look into installing Ansible via **Yum**, **Apt**, **Homebrew**, and **pip**.

Installing via Yum

If you are running a Fedora system, you can install Ansible directly, since from Fedora 22, Ansible 2.0+ is available in the official repositories. You can install it as follows:

```
$ sudo dnf install ansible
```

For RHEL and RHEL-based (CentOS, Scientific Linux, and Unbreakable Linux) systems, versions 6 and 7 have Ansible 2.0+ available in the EPEL repository, so you should ensure that you have the EPEL repository enabled before installing Ansible as follows:

```
$ sudo yum install ansible
```

 On RHEL 6, you have to run the `$ sudo rpm -Uvh http://dl.fedoraproject.org/pub/epel/6/x86_64/epel-release-6-8.noarch.rpm` command to install EPEL, while on RHEL 7, `$ sudo yum install epel-release` is enough.

Installing via Apt

Ansible is available for Ubuntu and Debian. To install Ansible on those operating systems, use the following command:

```
$ sudo apt-get install ansible
```

Installing via Homebrew

You can install Ansible on Mac OS X using Homebrew, as follows:

```
$ brew update
$ brew install ansible
```

Installing via pip

You can install Ansible via pip. If you don't have pip installed on your system, install it. You can use pip to install Ansible on Windows too, using the following command line:

```
$ sudo easy_install pip
```

You can now install Ansible using `pip`, as follows:

```
$ sudo pip install ansible
```

Once you're done installing Ansible, run `ansible --version` to verify that it has been installed:

```
$ ansible --version
```

You will get many lines as output from the preceding command line, as follows:

```
ansible 2.7.1
  config file = /etc/ansible/ansible.cfg
  configured module search path = [u'/home/fale/.ansible/plugins/modules',
u'/usr/share/ansible/plugins/modules']
  ansible python module location = /usr/lib/python2.7/site-packages/ansible
  executable location = /bin/ansible
  python version = 2.7.15 (default, Oct 15 2018, 15:24:06) [GCC 8.1.1
20180712 (Red Hat 8.1.1-5)]
```

Installing Ansible from source

In case the previous methods do not fit your use case, you can install Ansible directly from source. Installing from source does not require any root permissions. Let's clone a repository and activate `virtualenv`, which is an isolated environment in Python where you can install packages without interfering with the system's Python packages. The command and the resulting output for the repository are as follows:

```
$ git clone git://github.com/ansible/ansible.git
Cloning into 'ansible'...
remote: Counting objects: 116403, done.
remote: Compressing objects: 100% (18/18), done.
remote: Total 116403 (delta 3), reused 0 (delta 0), pack-reused 116384
Receiving objects: 100% (116403/116403), 40.80 MiB | 844.00 KiB/s,
done.
Resolving deltas: 100% (69450/69450), done.
Checking connectivity... done.
$ cd ansible/
$ source ./hacking/env-setup
Setting up Ansible to run out of checkout...
PATH=/home/vagrant/ansible/bin:/usr/local/bin:/bin:/usr/bin:/usr/local/sbin
:/usr/sbin:/sbin:/home/vagrant/bin
PYTHONPATH=/home/vagrant/ansible/lib:
MANPATH=/home/vagrant/ansible/docs/man:
Remember, you may wish to specify your host file with -i
Done!
```

Ansible needs a couple of Python packages, which you can install using `pip`. If you don't have pip installed on your system, install it using the following command. If you don't have `easy_install` installed, you can install it using Python's `setuptools` package on Red Hat systems, or by using Brew on the macOS:

```
$ sudo easy_install pip
<A long output follows>
```

Once you have installed `pip`, install the `paramiko`, `PyYAML`, `jinja2`, and `httplib2` packages using the following command lines:

```
$ sudo pip install paramiko PyYAML jinja2 httplib2
Requirement already satisfied (use --upgrade to upgrade): paramiko in
/usr/lib/python2.6/site-packages
Requirement already satisfied (use --upgrade to upgrade): PyYAML in
/usr/lib64/python2.6/site-packages
Requirement already satisfied (use --upgrade to upgrade): jinja2 in
/usr/lib/python2.6/site-packages
Requirement already satisfied (use --upgrade to upgrade): httplib2 in
/usr/lib/python2.6/site-packages
Downloading/unpacking markupsafe (from jinja2)
  Downloading MarkupSafe-0.23.tar.gz
  Running setup.py (path:/tmp/pip_build_root/markupsafe/setup.py)
egg_info for package markupsafe
Installing collected packages: markupsafe
  Running setup.py install for markupsafe
    building 'markupsafe._speedups' extension
    gcc -pthread -fno-strict-aliasing -O2 -g -pipe -Wall -Wp,-
D_FORTIFY_SOURCE=2 -fexceptions -fstack-protector --param=ssp-buffer-size=4
-m64 -mtune=generic -D_GNU_SOURCE -fPIC -fwrapv -DNDEBUG -O2 -g -pipe -Wall
-Wp,-D_FORTIFY_SOURCE=2 -fexceptions -fstack-protector --param=ssp-buffer-
size=4 -m64 -mtune=generic -D_GNU_SOURCE -fPIC -fwrapv -fPIC -
I/usr/include/python2.6 -c markupsafe/_speedups.c -o build/temp.linux-
x86_64-2.6/markupsafe/_speedups.o
    gcc -pthread -shared build/temp.linux-
x86_64-2.6/markupsafe/_speedups.o -L/usr/lib64 -lpython2.6 -o
build/lib.linux-x86_64-2.6/markupsafe/_speedups.so
Successfully installed markupsafe
Cleaning up...
```

 By default, Ansible will be running against the development branch. You might want to check out the latest stable branch. Check what the latest stable version is using the following `$ git branch -a` command.

Copy the latest version you want to use.

Version 2.0.2 was the latest version available at the time of writing. Check the latest version using the following command lines:

```
[node ansible]$ git checkout v2.7.1
Note: checking out 'v2.0.2'.
[node ansible]$ ansible --version
ansible 2.7.1 (v2.7.1 c963ef1dfb) last updated 2018/10/25 20:12:52 (GMT
+000)
```

You now have a working setup of Ansible ready. One of the benefits of running Ansible from source is that you can enjoy the new features immediately, without waiting for your package manager to make them available for you.

Creating a test environment with Vagrant

To be able to learn Ansible, we will need to make quite a few playbooks and run them.

Doing this directly on your computer will be very risky. For this reason, I would suggest using virtual machines.

It's possible to create a test environment with cloud providers in a few seconds, but it is often more useful to have those machines locally. To do so, we will use Vagrant, which is a piece of software by Hashicorp that allows users to quickly set up virtual environments independently from the virtualization backend used on the local system. It does support many virtualization backends (in the Vagrant ecosystem these are known as *Providers*) such as Hyper-V, VirtualBox, Docker, VMWare, and libvirt. This allows you to use the same syntax no matter what operating system or environment you are in.

First we will install `vagrant`. On Fedora, it will be enough to run the following code:

```
$ sudo dnf install -y vagrant
```

On Red Hat/CentOS/Scientific Linux/Unbreakable Linux, we will need to install `libvirt` first, enable it, and then install `vagrant` from the Hashicorp website:

```
$ sudo yum install -y qemu-kvm libvirt virt-install bridge-utils libvirt-
devel libxslt-devel libxml2-devel libvirt-devel libguestfs-tools-c
$ sudo systemctl enable libvirtd
$ sudo systemctl start libvirtd
$ sudo rpm -Uvh
https://releases.hashicorp.com/vagrant/2.2.1/vagrant_2.2.1_x86_64.rpm
$ vagrant plugin install vagrant-libvirt
```

If you use Ubuntu or Debian, you can install it using the following code:

```
$ sudo apt install virtualbox vagrant
```

For the following examples, I'll be virtualizing CentOS 7 machines. This is for multiple reasons; the main ones are as follows:

- CentOS is free and 100% compatible with Red Hat, Scientific Linux, and Unbreakable Linux.
- Many companies use Red Hat/CentOS/Scientific Linux/Unbreakable Linux for their servers.
- These distributions are the only ones with SELinux support built in, and, as we have seen earlier, SELinux can help you make your environment much more secure.

To test that everything went well, we can run the following commands:

```
$ sudo vagrant init centos/7 && sudo vagrant up
```

If everything went well, you should expect an output ending with something like this:

```
==> default: Configuring and enabling network interfaces...
    default: SSH address: 192.168.121.60:22
    default: SSH username: vagrant
    default: SSH auth method: private key
==> default: Rsyncing folder: /tmp/ch01/ => /vagrant
```

So, you can now execute `vagrant ssh`, and you will find yourself in the machine we just created.

 There will be a `Vagrant` file in the current folder. In this file, you can create the directives with `vagrant init` to create the virtual environment.

Version control systems

In this chapter, we have already encountered the expression **infrastructure code** to describe the Ansible code that will create and maintain your infrastructure. We use the expression infrastructure code to distinguish it from the application code, which is the code that composes your applications, websites, and so on. This distinction is needed for clarity, but in the end, both types are a bunch of text files that the software will be able to read and interpret.

For this reason, a version control system will help you a lot. Its main advantages are as follows:

- The ability to have multiple people working simultaneously on the same project.
- The ability to perform code-reviews in a simple way.
- The ability to have multiple branches for multiple environments (that is, dev, test, QA, staging, and production).
- The ability to track a change so that we know when it was introduced, and who introduced it. This makes it easier to understand why that piece of code is there, months or years later.

These advantages are provided to you by the majority of version control systems out there.

Version control systems can be divided into three major groups, based on the three different models that they can implement:

- Local data model
- Client-server model
- Distributed model

The first category, the local data model, is the oldest (circa 1972) approach and is used for very specific use cases. This model requires all users to share the same filesystem. Famous examples of it are the **Revision Control System (RCS)** and **Source Code Control System (SCCS)**.

The second category, the client-server model, arrived later (circa 1990) and tried to solve the limitations of the local data model, creating a server that respected the local data model and a set of clients that dealt with the server instead of with the repository itself. This additional layer allowed multiple developers to use local files and synchronize them with a centralized server. Famous examples of this approach are Apache **Subversion (SVN)**, and **Concurrent Versions System (CVS)**.

The third category, the distributed model, arrived at the beginning of the twenty-first century, and tried to solve the limitations of the client-server model. In fact, in the client-server model, you could work on the code offline, but you needed to be *online* to commit the changes. The distributed model allows you to handle everything on your local repository (like the local data model), and to merge different repositories on different machines in an easy way. In this new model, it's possible to perform all actions as in the client-server model, with the added benefits of being able to work completely offline as well as the ability to merge changes between peers without passing by the centralized server. Examples of this model are BitKeeper (proprietary software), Git, GNU Bazaar, and Mercurial.

There are some additional advantages that will be provided by only the distributed model, such as the following:

- The possibility of making commits, browsing history, and performing any other action even if the server is not available.
- Easier management of multiple branches for different environments.

When it comes to infrastructure code, we have to consider that the infrastructure that retains and manages your infrastructure code is frequently kept in the infrastructure code itself. This is a recursive situation that can create problems. A distributed version control system will prevent this problem.

As for the simplicity of managing multiple branches, even if this is not a hard rule, often distributed version control systems have much better merge handling than the other kinds of version control systems.

Using Ansible with Git

For the reasons that we have just seen, and because of its huge popularity, I suggest always using Git for your Ansible repositories.

There are a few suggestions that I always provide to the people I talk to, so that Ansible gets the best out of Git:

- **Create environment branches**: Creating environment branches, such as dev, prod, test, and stg, will allow you to easily keep track of the different environments and their respective update statuses. I often suggest keeping the master branch for the development environment, since I find that many people are used to pushing new changes directly to the master. If you use a master for a production environment, people can inadvertently push changes in the production environment when they wanted to push them in a development environment.
- **Always keep environment branches stable**: One of the big advantages of having environment branches is the possibility of destroying and recreating any environment from scratch at any given moment. This is only possible if your environment branches are in a stable (not broken) state.

- **Use feature branches**: Using different branches for specific long-development features (such as a refactor or some other big changes) will allow you to keep your day-to-day operations while your new feature is in the Git repository (so you'll not lose track of who did what and when they did it).

- **Push often**: I always suggest that people *push commits* as often as possible. This will make Git work as both a version control system and a backup system. I have seen laptops broken, lost, or stolen with days or weeks of un-pushed work on them far too often. Don't waste your time – push often. Also, by pushing often, you'll detect merge conflicts sooner, and conflicts are always easier to handle when they are detected early, instead of waiting for multiple changes.

- **Always deploy after you have made a change**: I have seen times when a developer has created a change in the infrastructure code, tested in the dev and test environments, pushed to the production branch, and then went to have lunch before deploying the changes in production. His lunch did not end well. One of his colleagues deployed the code to production inadvertently (he was trying to deploy a small change he had made in the meantime) and was not prepared to handle the other developer's deployment. The production infrastructure broke and they lost a lot of time figuring out how it was possible that such a small change (the one the person who made the deployment was aware of) created such a big mess.

- **Choose multiple small changes rather than a few huge changes**: Making small changes, whenever possible, will make debugging easier. Debugging an infrastructure is not very easy. There is no compiler that will allow you to see "obvious problems" (even though Ansible performs a syntax check of your code, no other test is performed), and the tools for finding something that is broken are not always as good as you would imagine. The infrastructure as a code paradigm is new, and tools are not yet as good as the ones for the application code.

- **Avoid binary files as much as possible**: I always suggest keeping your binaries outside your Git repository, whether it is an application code repository or an infrastructure code repository. In the application code example, I think it is important to keep your repository light (Git, as well as the majority of the version control systems, do not perform very well with binary blobs), while, for the infrastructure code example, it is vital because you'll be tempted to put a huge number of binary blobs in it, since very often it is easier to put a binary blob in the repository than to find a cleaner (and better) solution.

Summary

In this chapter, we have seen what IT automation is, its advantages and disadvantages, what kind of tools you can find, and how Ansible fits into this big picture. We have also seen how to install Ansible and how to create a Vagrant virtual machine. In the end, we analyzed the version control systems and spoke about the advantages Git brings to Ansible, if used properly.

In the next chapter, we will start looking at the infrastructure code that we mentioned in this chapter, without explaining exactly what it is and how to write it. We'll also see how to automate simple operations that you probably perform every single day, such as managing users, managing files, and file content.

Automating Simple Tasks

2

As mentioned in the previous chapter, Ansible can be used to both create and manage a whole infrastructure, as well as be integrated into an infrastructure that is already working.

In this chapter, we will cover the following topics:

- YAML
- Working with Playbook
- Ansible velocity
- Variables in Playbook
- Creating the Ansible User
- Configuring a basic server
- Installing and configuring a web server
- Publishing a website
- Jinja2 templates

First, we will talk about **YAML Ain't Markup Language (YAML)**, a human-readable data serialization language that is widely used in Ansible.

Technical Requirement

You can download all the files from this book's GitHub repository at: `https://github.com/PacktPublishing/Learning-Ansible-2.X-Third-Edition/tree/master/Chapter02`.

YAML

YAML, like many other data serialization languages (such as JSON), has very few, basic concepts:

- Declarations
- Lists
- Associative arrays

A declaration is very similar to a variable in any other language, which is as follows:

```
name: 'This is the name'
```

To create a list, we will have to use –:

```
- 'item1'
- 'item2'
- 'item3'
```

YAML uses indentation to logically divide parents from children. So, if we want to create associative arrays (also known as objects), we would just need to add an indentation:

```
item:
  name: TheName
  location: TheLocation
```

Obviously, we can mix those together as follows:

```
people:
  - name: Albert
    number: +1000000000
    country: USA
  - name: David
    number: +44000000000
    country: UK
```

Those are the basics of YAML. YAML can do much more, but for now, this will be enough.

Hello Ansible

As we have seen in the previous chapter, it is possible to use Ansible to automate simple tasks that you probably already perform daily.

Let's start by checking whether or not a remote machine is reachable; in other words, let's start by pinging a machine. The simplest way to do this is to run the following:

```
$ ansible all -i HOST, -m ping
```

Here, HOST is an IP address, the **Fully Qualified Domain Name** (**FQDN**), or an alias of a machine where you have SSH access (you can use a **Vagrant** host, as we have seen in the previous chapter).

> After HOST, the comma is mandatory, because otherwise, it would not be seen as a list, but as a string.

In this case, we have performed it against a virtual machine on our system:

```
$ ansible all -i test01.fale.io, -m ping
```

You should receive something like this as a result:

```
test01.fale.io | SUCCESS => {
    "changed": false,
    "ping": "pong"
}
```

Now, let's see what we did and why. Let's start from the Ansible help. To query it, we can use the following command:

```
$ ansible --help
```

To make it easier to read, we have removed all the output related to options that we have not used:

```
Usage: ansible <host-pattern> [options]

Options:
  -i INVENTORY, --inventory=INVENTORY, --inventory-file=INVENTORY
                        specify inventory host path or comma separated host
                        list. --inventory-file is deprecated
  -m MODULE_NAME, --module-name=MODULE_NAME
                        module name to execute (default=command)
```

So, what we did was as follows:

1. We invoked Ansible.
2. We instructed Ansible to run on all hosts.
3. We specified our inventory (also known as the list of the hosts).
4. We specified the module we wanted to run (ping).

Now that we can ping the server, let's try echo hello ansible!, as shown in the following command:

```
$ ansible all -i test01.fale.io, -m shell -a '/bin/echo hello ansible!'
```

You should receive something like this as a result:

```
test01.fale.io | CHANGED | rc=0 >>
hello ansible!
```

In this example, we used an additional option. Let's check the Ansible help to see what it does:

```
Usage: ansible <host-pattern> [options]
Options:
  -a MODULE_ARGS, --args=MODULE_ARGS
                        module arguments
```

As you may have guessed from the context and the name, the args options allow you to pass additional arguments to the module. Some modules (such as ping) do not support any arguments, while others (such as shell) will require arguments.

Working with playbooks

Playbooks are one of the core features of Ansible and tell Ansible what to execute. They are like a to-do list for Ansible that contains a list of tasks; each task internally links to a piece of code called a **module**. Playbooks are simple, human-readable YAML files, while modules are a piece of code that can be written in any language, with the condition that its output be in the JSON format. You can have multiple tasks listed in a playbook, and these tasks would be executed serially by Ansible. You can think of playbooks as an equivalent of manifests in Puppet, states in Salt, or cookbooks in Chef; they allow you to enter a list of tasks or commands you want to execute on your remote system.

Studying the anatomy of a playbook

Playbooks can have a list of remote hosts, user variables, tasks, handlers, and so on. You can also override most of the configuration settings through a playbook. Let's start looking at the anatomy of a playbook.

The purpose of the playbook that we are going to consider now is to ensure that the `httpd` package is installed and the service is **enabled** and **started**. This is the content of the `setup_apache.yaml` file:

```
- hosts: all
  remote_user: vagrant
  tasks:
    - name: Ensure the HTTPd package is installed
      yum:
        name: httpd
        state: present
      become: True
    - name: Ensure the HTTPd service is enabled and running
      service:
        name: httpd
        state: started
        enabled: True
      become: True
```

The `setup_apache.yaml` file is an example of a playbook. The file is comprised of three main parts, as follows:

- `hosts`: This lists the host or host group that we want to run the task against. The hosts field is required. It is used by Ansible to determine which hosts will be targeted by the listed tasks. If a host group is provided instead of a host, Ansible will try to look up the hosts belonging to it based on the inventory file. If there is no match, Ansible will skip all the tasks for that host group. The `--list-hosts` option, along with the playbook (`ansible-playbook <playbook> --list-hosts`), will tell you exactly which hosts the playbook will run against.

- `remote_user`: This is one of the configuration parameters of Ansible (consider, for example, `tom' - remote_user`) that tells Ansible to use a particular user (in this case, `tom`) while logging into the system.

- `tasks`: Finally, we come to tasks. All playbooks should contain tasks. Tasks are a list of actions that you want to perform. A `tasks` field contains the name of the task (that is, the help text for the user about the task), a module that should be executed, and arguments that are required for the module. Let's look at the single task that is listed in the playbook, as shown in the preceding snippet of code.

All examples in the book would be executed on CentOS, but the same set of examples with a few changes would work on other distributions as well.

In the preceding case, there are two tasks. The `name` parameter represents what the task is doing and is `present` mainly to improve readability, as we'll see during the playbook run. The `name` parameter is optional. The `yum` and `service` modules have their own set of parameters. Almost all modules have the `name` parameter (there are exceptions, such as the `debug` module), which indicates what component the actions are performed on. Let's look at the other parameters:

- The `state` parameter holds the latest value in the `yum` module, and it indicates that the `httpd` package should have been installed. The command to execute roughly translates to `yum install httpd`.
- In the `service` module's scenario, the `state` parameter with the started value indicates that the `httpd` service should be started, and it roughly translates to a `/etc/init.d/httpd` start. In this module, we also have the `enabled` parameter, which defines whether the service should start at boot or not.
- The `become: True` parameter represents the fact that the tasks should be executed with `sudo` access. If the `sudo` user's file does not allow the user to run the particular command, then the playbook will fail when it is run.

  You might have questions about why there is no package module that figures out the architecture internally and runs the `yum`, `apt`, or any other package options depending on the architecture of the system. Ansible populates the package manager value into a variable named `ansible_pkg_manager`.

 In general, we need to remember that the number of packages that have a common name across different operating systems are a small subset of the number of packages that are actually present. For example, the `httpd` package is called `httpd` in Red Hat systems, but is called `apache2` in Debian-based systems. We also need to remember that every package manager has its own set of options that make it powerful; as a result, it makes more sense to use explicit package manager names so that the full set of options are available to the end user writing the playbook.

Running a playbook

Now, it's time (yes, finally!) to run the playbook. To instruct Ansible to execute a playbook instead of a module, we will have to use a different command (`ansible-playbooks`) that has a syntax very similar to the `ansible` command we already saw:

```
$ ansible-playbook -i HOST, setup_apache.yaml
```

As you can see, aside from the host-pattern (that is specified in the playbook), which has disappeared, and the module option, which has been replaced by the playbook name, nothing has changed. So, to execute this command on my machine, the exact command is as follows:

```
$ ansible-playbook -i test01.fale.io, setup_apache.yaml
```

The result is the following:

```
PLAY [all] ***********************************************************

TASK [Gathering Facts] **********************************************
ok: [test01.fale.io]

TASK [Ensure the HTTPd package is installed] ***********************
changed: [test01.fale.io]

TASK [Ensure the HTTPd service is enabled and running] ************
changed: [test01.fale.io]

PLAY RECAP *********************************************************
test01.fale.io              : ok=3 changed=2 unreachable=0 failed=0
```

Wow! The example worked. Let's now check whether the `httpd` package is installed and is now up-and-running on the machine. To check if HTTPd is installed, the easiest way is to ask `rpm`:

```
$ rpm -qa | grep httpd
```

If everything worked properly, you should have an output like the following:

```
httpd-tools-2.4.6-80.el7.centos.1.x86_64
httpd-2.4.6-80.el7.centos.1.x86_64
```

To see the status of the service, we can ask `systemd`:

```
$ systemctl status httpd
```

The expected result is something like the following:

```
httpd.service - The Apache HTTP Server
   Loaded: loaded (/usr/lib/systemd/system/httpd.service; enabled; vendor
preset: disabled)
   Active: active (running) since Tue 2018-12-04 15:11:03 UTC; 29min ago
     Docs: man:httpd(8)
           man:apachectl(8)
 Main PID: 604 (httpd)
   Status: "Total requests: 0; Current requests/sec: 0; Current traffic: 0
B/sec"
   CGroup: /system.slice/httpd.service
           ├─604 /usr/sbin/httpd -DFOREGROUND
           ├─624 /usr/sbin/httpd -DFOREGROUND
           ├─626 /usr/sbin/httpd -DFOREGROUND
           ├─627 /usr/sbin/httpd -DFOREGROUND
           ├─628 /usr/sbin/httpd -DFOREGROUND
           └─629 /usr/sbin/httpd -DFOREGROUND
```

The end state, according to the playbook, has been achieved. Let's briefly look at exactly what happens during the playbook run:

```
PLAY [all] ********************************************************
```

This line advises us that a playbook is going to start here, and that it will be executed on all hosts:

```
TASK [Gathering Facts] *******************************************
ok: [test01.fale.io]
```

The TASK lines show the name of the task (setup, in this case), and its effect on each host. Sometimes, people get confused by the setup task. In fact, if you look at the playbook, there is no setup task. This is because Ansible, before executing the tasks that we have asked it to, will try to connect to the machine and gather information about it that could be useful later. As you can see, the task resulted in a green ok state, so it succeeded, and nothing was changed on the server:

```
TASK [Ensure the HTTPd package is installed] *************************
changed: [test01.fale.io]

TASK [Ensure the HTTPd service is enabled and running] **************
changed: [test01.fale.io]
```

These two task's states are yellow, and spell `changed`. This means that those tasks were executed and have succeeded, but they have actually changed something on the machine:

```
PLAY RECAP ***********************************************************
test01.fale.io : ok=3 changed=2 unreachable=0 failed=0
```

Those last few lines are a recapitulation of how the playbook went. Let's rerun the task now and see the output after both the tasks have actually run:

```
PLAY [all] ***********************************************************

TASK [Gathering Facts] **********************************************
ok: [test01.fale.io]

TASK [Ensure the HTTPd package is installed] ***********************
ok: [test01.fale.io]

TASK [Ensure the HTTPd service is enabled and running] ************
ok: [test01.fale.io]

PLAY RECAP ***********************************************************
test01.fale.io                : ok=3 changed=0 unreachable=0 failed=0
```

As you would have expected, the two tasks in question give an output of `ok`, which means that the desired state was already met prior to running the task. It's important to remember that many tasks, such as the **gathering facts** task, obtain information regarding a particular component of the system and do not necessarily change anything on the system; so, these tasks didn't display the changed output earlier.

The PLAY RECAP section in the first and second run are shown as follows. You will see the following output during the first run:

```
PLAY RECAP ***********************************************************
test01.fale.io                : ok=3 changed=2 unreachable=0 failed=0
```

You will see the following output during the second run:

```
PLAY RECAP ***********************************************************
test01.fale.io                : ok=3 changed=0 unreachable=0 failed=0
```

As you can see, the difference is that the first task's output shows changed=2, which means that the system state changed twice due to two tasks. It's very useful to look at this output, since, if a system has achieved its desired state and then you run the playbook on it, the expected output should be changed=0.

If you're thinking of the word **idempotency** at this stage, you're absolutely right and deserve a pat on the back! Idempotency is one of the key tenets of configuration management. Wikipedia defines idempotency as an operation that, if applied twice to any value, gives the same result as if it were applied once. The earliest examples of this that you would have encountered in your childhood would be multiplicative operations on the number 1, where 1*1=1 every single time.

Most of the configuration management tools have taken this principle and applied it to the infrastructure as well. In a large infrastructure, it is highly recommended to monitor or track the number of changed tasks in your infrastructure and alert the concerned tasks if you find oddities; this applies to any configuration management tool in general. In an ideal state, the only time you should see changes is when you're introducing a new change in the form of any **Create**, **Remove**, **Update**, or **Delete** (**CRUD**) operation on various system components. If you're wondering how you can do it with Ansible, keep reading the book and you'll eventually find the answer!

Let's proceed. You could have also written the preceding tasks as follows, but when the tasks are run from an end user's perspective, they are quite readable (we will call this file `setup_apache_no_com.yaml`):

```
---
- hosts: all
  remote_user: vagrant
  tasks:
    - yum:
        name: httpd
        state: present
      become: True
    - service:
        name: httpd
        state: started
        enabled: True
      become: True
```

Let's run the playbook again to spot any difference in the output:

```
$ ansible-playbook -i test01.fale.io, setup_apache_no_com.yaml
```

The output would be as follows:

```
PLAY [all] ************************************************************

TASK [Gathering Facts] ***********************************************
ok: [test01.fale.io]

TASK [yum] ***********************************************************
```

```
ok: [test01.fale.io]

TASK [service] ***********************************************
ok: [test01.fale.io]

PLAY RECAP **************************************************
test01.fale.io              : ok=3 changed=0 unreachable=0 failed=0
```

As you can see, the difference is in the readability. Wherever possible, it's recommended to keep the tasks as simple as possible (the **KISS** principle: **Keep It Simple, Stupid**) to allow for the maintainability of your scripts in the long run.

Now that we've seen how you can write a basic playbook and run it against a host, let's look at other options that would help you while running playbooks.

Ansible verbosity

One of the first options that anyone picks up is the debug option. To understand what is happening when you run the playbook, you can run it with the verbose (–v) option. Every extra v will provide the end user with more debug output.

Let's see an example of using those options to debug a simple `ping` command (`ansible all -i test01.fale.io, -m ping`):

- The –v option provides the default output:

```
Using /etc/ansible/ansible.cfg as config file
test01.fale.io | SUCCESS => {
    "changed": false,
    "ping": "pong"
}
```

- The –vv option adds a little more information about the Ansible environment and the handlers:

```
ansible 2.7.2
  config file = /etc/ansible/ansible.cfg
  configured module search path =
[u'/home/fale/.ansible/plugins/modules',
u'/usr/share/ansible/plugins/modules']
  ansible python module location = /usr/lib/python2.7/site-
packages/ansible
  executable location = /bin/ansible
  python version = 2.7.15 (default, Oct 15 2018, 15:24:06) [GCC
8.1.1 20180712 (Red Hat 8.1.1-5)]
```

```
Using /etc/ansible/ansible.cfg as config file
META: ran handlers
test01.fale.io | SUCCESS => {
    "changed": false,
    "ping": "pong"
}
META: ran handlers
META: ran handlers
```

- The –vvv option adds a lot more information. For instance, it shows the ssh command that Ansible uses to create a temporary file on the remote host and run the script remotely. Full script is available on GitHub.

```
ansible 2.7.2
  config file = /etc/ansible/ansible.cfg
  configured module search path =
[u'/home/fale/.ansible/plugins/modules',
u'/usr/share/ansible/plugins/modules']
  ansible python module location = /usr/lib/python2.7/site-
packages/ansible
  executable location = /bin/ansible
  python version = 2.7.15 (default, Oct 15 2018, 15:24:06) [GCC
8.1.1 20180712 (Red Hat 8.1.1-5)]
Using /etc/ansible/ansible.cfg as config file
Parsed test01.fale.io, inventory source with host_list plugin
META: ran handlers
<test01.fale.io> ESTABLISH SSH CONNECTION FOR USER: None
<test01.fale.io> SSH: EXEC ssh -C -o ControlMaster=auto -o
...
```

Now we understood what is happening when you run the playbook, with the verbose -vvv option.

Variables in playbooks

Sometimes, it is important to set and get variables in a playbook.

Very often, you'll need to automate multiple similar operations. In those cases, you'll want to create a single playbook that can be called with different variables to ensure code reusability.

Another case where variables are very important is when you have more than one data center, and some values will be data center-specific. A common example are the DNS servers. Let's analyze the following simple code that will introduce us to the Ansible way to set and get variables:

```
- hosts: all
  remote_user: vagrant
  tasks:
    - name: Set variable 'name'
      set_fact:
        name: Test machine
    - name: Print variable 'name'
      debug:
        msg: '{{ name }}'
```

Let's run it in the usual way:

```
$ ansible-playbook -i test01.fale.io, variables.yaml
```

You should see the following result:

```
PLAY [all] ***********************************************************

TASK [Gathering Facts] ***********************************************
ok: [test01.fale.io]

TASK [Set variable 'name'] *******************************************
ok: [test01.fale.io]

TASK [Print variable 'name'] *****************************************
ok: [test01.fale.io] => {
    "msg": "Test machine"
}

PLAY RECAP ***********************************************************
test01.fale.io             : ok=3 changed=0 unreachable=0 failed=0
```

If we analyze the code we have just executed, it should be pretty clear what's going on. We set a variable (that in Ansible are called facts) and then we print it with the debug function.

Variables should always be between quotes when you use this expanded version of YAML.

Ansible allows you to set your variables in many different ways – that is, either by passing a variable file, declaring it in a playbook, passing it to the `ansible-playbook` command using `-e` / `--extra-vars`, or by declaring it in an inventory file (we will be discussing this in depth in the next chapter).

It's now time to start using some metadata that Ansible obtained during the setup phase. Let's start by looking at the data that is gathered by Ansible. To do this, we will execute the following code:

```
$ ansible all -i HOST, -m setup
```

In our specific case, this means executing the following code:

```
$ ansible all -i test01.fale.io, -m setup
```

We can obviously do the same with a playbook, but this way is faster. Also, for the `setup` case, you will only need to see the output during the development to be sure to use the right variable name for your goal.

The output will be something like this. Full code output is available on GitHub.

```
test01.fale.io | SUCCESS => {
    "ansible_facts": {
        "ansible_all_ipv4_addresses": [
            "192.168.121.190"
        ],
        "ansible_all_ipv6_addresses": [
            "fe80::5054:ff:fe93:f113"
        ],
        "ansible_apparmor": {
            "status": "disabled"
        },
        "ansible_architecture": "x86_64",
        "ansible_bios_date": "04/01/2014",
        "ansible_bios_version": "?-20180531_142017-
buildhw-08.phx2.fedoraproject.org-1.fc28",
        ...
```

As you can see from this huge list of options, you can gain a huge quantity of information, and you can use them as any other variable. Let's print the OS name and the version. To do so, we can create a new playbook called `setup_variables.yaml`, with the following content:

```
- hosts: all
  remote_user: vagrant
  tasks:
    - name: Print OS and version
```

```
    debug:
        msg: '{{ ansible_distribution }} {{ ansible_distribution_version
}}'
```

Run it with the following code:

```
$ ansible-playbook -i test01.fale.io, setup_variables.yaml
```

This will give us the following output:

```
PLAY [all] *********************************************************

TASK [Gathering Facts] ********************************************
ok: [test01.fale.io]

TASK [Print OS and version] **************************************
ok: [test01.fale.io] => {
    "msg": "CentOS 7.5.1804"
}

PLAY RECAP *******************************************************
test01.fale.io              : ok=2 changed=0 unreachable=0 failed=0
```

As you can see, it printed the OS name and version as expected. In addition to the methods seen previously, it's also possible to pass a variable using a command-line argument. In fact, if we look in the Ansible help, we will notice the following:

```
Usage: ansible <host-pattern> [options]

Options:
  -e EXTRA_VARS, --extra-vars=EXTRA_VARS
                          set additional variables as key=value or YAML/JSON,
if
                          filename prepend with @
```

The same lines are present in the `ansible-playbook` command, as well. Let's make a small playbook called `cli_variables.yaml`, with the following content:

```
---
- hosts: all
  remote_user: vagrant
  tasks:
    - name: Print variable 'name'
      debug:
        msg: '{{ name }}'
```

Execute it with the following:

```
$ ansible-playbook -i test01.fale.io, cli_variables.yaml -e 'name=test01'
```

We will receive the following:

```
 [WARNING]: Found variable using reserved name: name

PLAY [all] *********************************************************

TASK [Gathering Facts] ********************************************
ok: [test01.fale.io]

TASK [Print variable 'name'] *************************************
ok: [test01.fale.io] => {
    "msg": "test01"
}

PLAY RECAP ********************************************************
test01.fale.io              : ok=2 changed=0 unreachable=0 failed=0
```

If we forgot to add the additional parameter to specify the variable, we would have executed it as follows:

```
$ ansible-playbook -i test01.fale.io, cli_variables.yaml
```

We would have received the following output:

```
PLAY [all] *********************************************************

TASK [Gathering Facts] ********************************************
ok: [test01.fale.io]

TASK [Print variable 'name'] *************************************
fatal: [test01.fale.io]: FAILED! => {"msg": "The task includes an option
with an undefined variable. The error was: 'name' is undefined\n\nThe error
appears to have been in '/home/fale/Learning-Ansible-2.X-Third-
Edition/Ch2/cli_variables.yaml': line 5, column 7, but may\nbe elsewhere in
the file depending on the exact syntax problem.\n\nThe offending line
appears to be:\n\n tasks:\n - name: Print variable 'name'\n ^ here\n"}
 to retry, use: --limit @/home/fale/Learning-Ansible-2.X-Third-
Edition/Ch2/cli_variables.retry

PLAY RECAP ********************************************************
test01.fale.io : ok=1 changed=0 unreachable=0 failed=1
```

Now that we have learned the basics of playbooks, let's create a web server from scratch using them. To do so, let's start from the beginning, creating an Ansible user and then moving forward from there.

> As you will notice in the previous example, a **WARNING** popped up, informing us that we were re-declaring a reserved variable (name). The full list of reserved variables (as of Ansible 2.7) are as follows: `add,`
> `append, as_integer_ratio, bit_length, capitalize, center,`
> `clear, conjugate, copy, count, decode, denominator, difference,`
> `difference_update, discard, encode, endswith, expandtabs,`
> `extend, find, format, fromhex, fromkeys, get, has_key, hex, imag,`
> `index, insert, intersection, intersection_update, isalnum,`
> `isalpha, isdecimal, isdigit, isdisjoint, is_integer, islower,`
> `isnumeric, isspace, issubset, issuperset, istitle, isupper,`
> `items, iteritems, iterkeys, itervalues, join, keys, ljust, lower,`
> `lstrip, numerator, partition, pop, popitem, real, remove, replace,`
> `reverse, rfind, rindex, rjust, rpartition, rsplit, rstrip,`
> `setdefault, sort, split, splitlines, startswith, strip, swapcase,`
> `symmetric_difference, symmetric_difference_update, title,`
> `translate, union, update, upper, values, viewitems, viewkeys,`
> `viewvalues, zfill.`

Creating the Ansible user

When you create a machine (or rent one from any hosting company), it arrives with only the `root` user, or other users such as `vagrant`. Let's start creating a playbook that ensures that an Ansible user is created, it's accessible with an SSH key, and is able to perform actions on behalf of other users (`sudo`) with no password required. We often call this playbook `firstrun.yaml`, since we execute it as soon as a new machine is created, but after that, we don't use it, since we disable the default user for security reasons. Our script will look something like the following:

```
---
- hosts: all
  user: vagrant
  tasks:
    - name: Ensure ansible user exists
      user:
        name: ansible
        state: present
        comment: Ansible
```

```
        become: True
    - name: Ensure ansible user accepts the SSH key
      authorized_key:
        user: ansible
        key: https://github.com/fale.keys
        state: present
      become: True
    - name: Ensure the ansible user is sudoer with no password required
      lineinfile:
        dest: /etc/sudoers
        state: present
        regexp: '^ansible ALL\='
        line: 'ansible ALL=(ALL) NOPASSWD:ALL'
        validate: 'visudo -cf %s'
      become: True
```

Before running it, let's look at it a little bit. We have used three different modules (user, authorized_key, and lineinfile) that we have never seen.

The user module, as the name suggests, allows us to make sure a user is present (or absent).

The authorized_key module allows us to ensure that a certain SSH key can be used to log in as a specific user on that machine. This module will not substitute all the SSH keys that are already enabled for that user, but will simply add (or remove) the specified key. If you want to alter this behavior, you can use the exclusive option, which allows you to delete all the SSH keys that are not specified in this step.

The lineinfile module allows us to alter the content of a file. It works in a very similar way to **sed** (a stream editor), where you specify the regular expression that will be used to match the line, and then specify the new line that will be used to substitute the matched line. If no line is matched, the line is added at the end of the file.

Now let's run it with the following code:

```
$ ansible-playbook -i test01.fale.io, firstrun.yaml
```

This will give us the following result:

```
PLAY [all] *********************************************************

TASK [Gathering Facts] ********************************************
ok: [test01.fale.io]

TASK [Ensure ansible user exists] ********************************
changed: [test01.fale.io]
```

```
TASK [Ensure ansible user accepts the SSH key] ********************
changed: [test01.fale.io]

TASK [Ensure the ansible user is sudoer with no password required] *
changed: [test01.fale.io]

PLAY RECAP ********************************************************
test01.fale.io            : ok=4 changed=3 unreachable=0 failed=0
```

Configuring a basic server

After we have created the user for Ansible with the necessary privileges, we can go on to make some other small changes to the OS. To make it clearer, we will see how each action is performed, and then we'll look at the whole playbook.

Enabling EPEL

EPEL is the most important repository for Enterprise Linux and it contains a lot of additional packages. It's also a safe repository, since no package in EPEL will conflict with packages in the base repository.

To enable EPEL in RHEL/CentOS 7, it is enough to just install the `epel-release` package. To do so in Ansible, we will use the following:

```
- name: Ensure EPEL is enabled
  yum:
    name: epel-release
    state: present
  become: True
```

As you can see, we have used the `yum` module, as we did in one of the first examples of the chapter, specifying the name of the package and that we want it to be present.

Installing Python bindings for SELinux

Since Ansible is written in Python, and mainly uses the Python bindings to operate on the operating system, we will need to install the Python bindings for SELinux:

```
- name: Ensure libselinux-python is present
  yum:
    name: libselinux-python
```

```
      state: present
    become: True
  - name: Ensure libsemanage-python is present
    yum:
      name: libsemanage-python
      state: present
    become: True
```

 This could be written in a shorter way, using a cycle, but we'll see how to do this in the next chapter.

Upgrading all installed packages

To upgrade all installed packages, we will need to use the yum module again, but with a different parameter; in fact, we would use the following:

```
  - name: Ensure we have last version of every package
    yum:
      name: "*"
      state: latest
    become: True
```

As you can see, we have specified * as the package name (this stands for a wildcard to match all installed packages) and the state parameter is latest. This will upgrade all installed packages to the latest version available.

You may remember that, when we talked about the present state, we said that it was going to install the last available version. So, what's the difference between present and latest? present will install the latest version if the package is not installed, while if the package is already installed (no matter the version), it will go forward without making any change. latest will install the latest version if the package is not installed, and if the package is already installed, it will check whether a newer version is available, and, if it is, Ansible will update the package.

Ensuring that NTP is installed, configured, and running

To make sure NTP is present, we use the `yum` module:

```
- name: Ensure NTP is installed
  yum:
    name: ntp
    state: present
  become: True
```

Now that we know that NTP is installed, we should ensure that the server is using the `timezone` that we want. To do so, we will create a symbolic link in `/etc/localtime` that will point to the wanted `zoneinfo` file:

```
- name: Ensure the timezone is set to UTC
  file:
    src: /usr/share/zoneinfo/GMT
    dest: /etc/localtime
    state: link
  become: True
```

As you can see, we have used the `file` module, specifying that it needs to be a link (`state: link`).

To complete the NTP configuration, we need to start the `ntpd` service and ensure that it will run at every consequent boot:

```
- name: Ensure the NTP service is running and enabled
  service:
    name: ntpd
    state: started
    enabled: True
  become: True
```

Ensuring that FirewallD is present and enabled

As you can imagine, the first step is to ensure that FirewallD is installed:

```
- name: Ensure FirewallD is installed
  yum:
    name: firewalld
    state: present
  become: True
```

Since we want to be sure that we will not lose our SSH connection when we enable FirewallD, we will ensure that SSH traffic can always pass through it:

```
- name: Ensure SSH can pass the firewall
  firewalld:
    service: ssh
    state: enabled
    permanent: True
    immediate: True
  become: True
```

To do so, we have used the `firewalld` module. This module will take parameters that are very similar to the ones the `firewall-cmd` console would use. You will have to specify the service that is to be authorized to pass the firewall, whether you want this rule to apply immediately or not, and whether or not you want the rule to be permanent, so that after a reboot the rule will still be present.

> You can specify the service name (such as `ssh`) using the `service` parameter, or you can specify the port (such as `22/tcp`) using the `port` parameter.

Now that we have installed FirewallD and we are sure that our SSH connection will survive, we can enable it as we do any other service:

```
- name: Ensure FirewallD is running
  service:
    name: firewalld
    state: started
    enabled: True
  become: True
```

Adding a customized MOTD

To add the MOTD, we will need a template that will be the same for all servers, and a task to use the template.

I find it very useful to add a MOTD to every server. It's even more useful if you use Ansible, because you can use it to warn your users that changes to the system could be overwritten by Ansible. My usual template is called `motd`, and has this content:

```
               This system is managed by Ansible
     Any change done on this system could be overwritten by Ansible
OS: {{ ansible_distribution }} {{ ansible_distribution_version }}
Hostname: {{ inventory_hostname }}
eth0 address: {{ ansible_eth0.ipv4.address }}

            All connections are monitored and recorded
       Disconnect IMMEDIATELY if you are not an authorized user
```

This is a `jinja2` template, and it allows us to use every variable set in the playbooks. This also allows us to use complex syntax for conditionals and cycles that we will see later in this chapter. To populate a file from a template in Ansible, we will need to use the following:

```
- name: Ensure the MOTD file is present and updated
  template:
    src: motd
    dest: /etc/motd
    owner: root
    group: root
    mode: 0644
  become: True
```

The `template` module allows us to specify a local file (`src`) that will be interpreted by `jinja2`, and the output of this operation will be saved on the remote machine in a specific path (`dest`), will be owned by a specific user (`owner`) and group (`group`), and will have a specific access mode (`mode`).

Changing the hostname

To keep things simple, I find it useful to set the hostname of a machine to something meaningful. To do so, we can use a very simple Ansible module called `hostname`:

```
- name: Ensure the hostname is the same of the inventory
  hostname:
    name: "{{ inventory_hostname }}"
  become: True
```

Reviewing and running the playbook

Putting everything together, we now have the following playbook (called `common_tasks.yaml` for simplicity):

```yaml
---
- hosts: all
  remote_user: ansible
  tasks:
    - name: Ensure EPEL is enabled
      yum:
        name: epel-release
        state: present
      become: True
    - name: Ensure libselinux-python is present
      yum:
        name: libselinux-python
        state: present
      become: True
...
```

Since this `playbook` is pretty complex, we can run the following:

```
$ ansible-playbook common_tasks.yaml --list-tasks
```

This asks Ansible to print all the tasks in a shorter form so that we can quickly see what tasks a `playbook` performs. The output should be something like the following:

```
playbook: common_tasks.yaml
  play #1 (all): all TAGS: []
    tasks:
      Ensure EPEL is enabled TAGS: []
      Ensure libselinux-python is present TAGS: []
      Ensure libsemanage-python is present TAGS: []
      Ensure we have last version of every package TAGS: []
      Ensure NTP is installed TAGS: []
      Ensure the timezone is set to UTC TAGS: []
      Ensure the NTP service is running and enabled TAGS: []
      Ensure FirewallD is installed TAGS: []
      Ensure FirewallD is running TAGS: []
      Ensure SSH can pass the firewall TAGS: []
      Ensure the MOTD file is present and updated TAGS: []
      Ensure the hostname is the same of the inventory TAGS: []
```

We can now run the `playbook` with the following:

```
$ ansible-playbook -i test01.fale.io, common_tasks.yaml
```

We will receive the following output. Full code output is available on GitHub.

```
PLAY [all] ***************************************************

TASK [Gathering Facts] **************************************
ok: [test01.fale.io]

TASK [Ensure EPEL is enabled] ******************************
changed: [test01.fale.io]

TASK [Ensure libselinux-python is present] ****************
ok: [test01.fale.io]

TASK [Ensure libsemanage-python is present] **************
changed: [test01.fale.io]

TASK [Ensure we have last version of every package] *********
changed: [test01.fale.io]
...
```

Installing and configuring a web server

Now that we have made some generic changes to the operating system, let's move on to actually creating a web server. We are splitting those two phases so that we can share the first phase between every machine and apply the second only to web servers.

For this second phase, we will create a new playbook called `webserver.yaml` with the following content:

```
---
- hosts: all
  remote_user: ansible
  tasks:
    - name: Ensure the HTTPd package is installed
      yum:
        name: httpd
        state: present
      become: True
    - name: Ensure the HTTPd service is enabled and running
      service:
        name: httpd
        state: started
        enabled: True
      become: True
    - name: Ensure HTTP can pass the firewall
```

```
        firewalld:
          service: http
          state: enabled
          permanent: True
          immediate: True
        become: True
      - name: Ensure HTTPS can pass the firewall
        firewalld:
          service: https
          state: enabled
          permanent: True
          immediate: True
        become: True
```

As you can see, the first two tasks are the same as the ones in the example at the beginning of this chapter, and the last two tasks are used to instruct FirewallD to let the HTTP and HTTPS traffic pass.

Let's run this script with the following:

```
$ ansible-playbook -i test01.fale.io, webserver.yaml
```

This results in the following:

```
PLAY [all] ***************************************************

TASK [Gathering Facts] **************************************
ok: [test01.fale.io]

TASK [Ensure the HTTPd package is installed] **************
ok: [test01.fale.io]

TASK [Ensure the HTTPd service is enabled and running] *****
ok: [test01.fale.io]

TASK [Ensure HTTP can pass the firewall] ******************
changed: [test01.fale.io]

TASK [Ensure HTTPS can pass the firewall] *****************
changed: [test01.fale.io]

PLAY RECAP ************************************************
test01.fale.io      : ok=5 changed=2 unreachable=0 failed=0
```

Now that we have a web server, let's publish a small, single-page, static website.

Publishing a website

Since our website will be a simple, single-page website, we can easily create it and publish it using a single Ansible task. To make this page a little bit more interesting, we will create it from a template that will be populated by Ansible with a little data about the machine. The script to publish it will be called `deploy_website.yaml`, and will have the following content:

```
---
- hosts: all
  remote_user: ansible
  tasks:
    - name: Ensure the website is present and updated
      template:
        src: index.html.j2
        dest: /var/www/html/index.html
        owner: root
        group: root
        mode: 0644
      become: True
```

Let's start with a simple template that we will call `index.html.j2`:

```
<html>
    <body>
        <h1>Hello World!</h1>
    </body>
</html>
```

Now, we can test our website deployment by running the following:

```
$ ansible-playbook -i test01.fale.io, deploy_website.yaml
```

We should receive the following output:

```
PLAY [all] *********************************************

TASK [Gathering Facts] ********************************
ok: [test01.fale.io]

TASK [Ensure the website is present and updated] *********
changed: [test01.fale.io]

PLAY RECAP *******************************************
test01.fale.io    : ok=2 changed=1 unreachable=0 failed=0
```

If you now go to your IP/FQDN test machine in your browser, you'll find the **Hello World!** page.

Jinja2 templates

Jinja2 is a widely-used and fully-featured template engine for Python. Let's look at some syntax that will help us with Ansible. This paragraph is not a replacement for the official documentation, but its goal is to teach you some components that you'll find very useful when using with Ansible.

Variables

As we have seen, we can print variable content simply by using the `{{ VARIABLE_NAME }}` syntax. If we want to print just an element of an array, we can use `{{ ARRAY_NAME['KEY'] }}`, and if we want to print a property of an object, we can use `{{ OBJECT_NAME.PROPERTY_NAME }}`.

So, we can improve our previous static page in the following way:

```
<html>
    <body>
        <h1>Hello World!</h1>
        <p>This page was created on {{ ansible_date_time.date }}.</p>
    </body>
</html>
```

Filters

From time to time, we may want to change the style of a string a little bit, without writing specific code for it; for example, we may want to capitalize some text. To do so, we can use one of Jinja2's filters, such as `{{ VARIABLE_NAME | capitalize }}`. There are many filters available for Jinja2, and you can find the full list at `http://jinja.pocoo.org/docs/dev/templates/#builtin-filters`.

Conditionals

One thing you may often find useful in a template engine is the possibility of printing different strings depending on the content (or existence) of a string. So, we can improve our static web page in the following way:

```html
<html>
    <body>
        <h1>Hello World!</h1>
        <p>This page was created on {{ ansible_date_time.date }}.</p>
{% if ansible_eth0.active == True %}
        <p>eth0 address {{ ansible_eth0.ipv4.address }}.</p>
{% endif %}
    </body>
</html>
```

As you can see, we have added the capability to print the main IPv4 address for the `eth0` connection, if the connection is `active`. With conditionals, we can also use the tests.

> For a full list of builtin tests, please refer
> to `http://jinja.pocoo.org/docs/dev/templates/#builtin-tests`.

So, to obtain the same result, we could also have written the following:

```html
<html>
    <body>
        <h1>Hello World!</h1>
        <p>This page was created on {{ ansible_date_time.date }}.</p>
{% if ansible_eth0.active is equalto True %}
        <p>eth0 address {{ ansible_eth0.ipv4.address }}.</p>
{% endif %}
    </body>
</html>
```

There are a lot of different tests that will really help you to create easy-to-read, effective templates.

Cycles

The jinja2 template system also offers the capability to create cycles. Let's add a feature to our page that will print the main IPv4 network address for each device, instead of only eth0. We will then have the following code:

```
<html>
    <body>
        <h1>Hello World!</h1>
        <p>This page was created on {{ ansible_date_time.date }}.</p>
        <p>This machine can be reached on the following IP addresses</p>
        <ul>
{% for address in ansible_all_ipv4_addresses %}
            <li>{{ address }}</li>
{% endfor %}
        </ul>
    </body>
</html>
```

As you can see, the syntax for cycles is familiar if you already know Python.

These few pages on Jinja2 templating were not a substitute for the official documentation. In fact, Jinja2 templates are much more powerful than what we have seen here. The goal here is to give you the basic Jinja2 templates that are most often used in Ansible.

Summary

In this chapter, we started looking at YAML, and saw what a playbook is, how it works, and how to use it to create a web server (and a deployment for your static website). We have also seen multiple Ansible modules, such as the user, yum, service, FirewalID, lineinfile, and template modules. At the end of the chapter, we focused on templates.

In the next chapter, we will talk about inventories, so that we can easily manage multiple machines.

Section 2: Deploying Playbooks in a Production Environment

2

This section will help you to create deployments that have multiple phases and multiple machines. It will also explain how Ansible can be integrated with various cloud offerings and how it can simplify your life by managing the cloud for you.

This section contains the following chapters:

- Chapter 3, *Scaling to Multiple Hosts*
- Chapter 4, *Handling Complex Deployment*
- Chapter 5, *Going Cloud*
- Chapter 6, *Getting Notification from Ansible*

3
Scaling to Multiple Hosts

In the previous chapters, we have specified the hosts in the command line. This worked well while having a single host to work on, but it will not work very well when managing multiple servers. In this chapter, we will see exactly how to leverage inventories to manage multiple servers. Also, we will introduce topics such as host variables and group variables to make it possible to easily and quickly set up similar but different hosts. We will speak about loops in **Ansible**, which allows you to reduce the amount of code you write and at the same time make it more readable as well.

In this chapter, we will be covering the following topics:

- Working with inventory files
- Working with variables

Technical requirements

You can download all the files from this book's GitHub repository at `https://github.com/PacktPublishing/Learning-Ansible-2.X-Third-Edition/tree/master/Chapter03`.

Working with inventory files

An **inventory file** is the source of truth for Ansible (there is also an advanced concept called **dynamic inventory**, which we will cover later). It follows the **Initialization** (INI) format and tells Ansible whether the remote host or hosts provided by the user are genuine.

Ansible can run its tasks against multiple hosts in parallel. To do this, you can directly pass the list of hosts to Ansible using an inventory file. For such parallel execution, Ansible allows you to group your hosts in the inventory file; the file passes the group's name to Ansible. Ansible will search for that group in the inventory file and run its tasks against all the hosts listed in that group.

You can pass the inventory file to Ansible using the -i or --inventory-file option followed by the path to the file. If you do not explicitly specify any inventory file to Ansible, it will take the default path from the host_file parameter of ansible.cfg, which defaults to /etc/ansible/hosts.

> When using the -i parameter, if the value is a list (it contains at least one comma), it will be used as the inventory list, while if the variable is a string, it will be used as the inventory file path.

The basic inventory file

Before diving into the concept, let's first look at a basic inventory file called hosts that we can use instead of the list we used in the previous examples:

test01.fale.io

> Ansible can take either an FQDN or an IP address within the inventory file.

We can now perform the same operations that we did in the previous chapter, tweaking the Ansible command parameters.

For instance, to install the web server, we used this command:

```
$ ansible-playbook -i test01.fale.io, webserver.yaml
```

Instead, we can use the following:

```
$ ansible-playbook -i hosts webserver.yaml
```

As you can see, we have substituted the list of hosts with the inventory filename.

Groups in an inventory file

The advantages of inventory files are noticeable when we have more complex situations. Let's say our website is getting more complicated and we now need a more complex environment. In our example, our website will require a MySQL database. Also, we will decide to have two web servers. In this scenario, it makes sense to group different machines based on their role in our infrastructure.

Ansible allows us to create an INI-like file with groups (INI sections) and hosts. Here's what our hosts file would change to:

```
[webserver]
ws01.fale.io
ws02.fale.io

[database]
db01.fale.io
```

Now we can instruct playbooks to run only on hosts in a certain group. In the previous chapter, we have created three different playbooks for our website example:

- `firstrun.yaml` is generic and will have to be run on every machine.
- `common_tasks.yaml` is generic and will have to be run on every machine.
- `webserver.yaml` is specific for web servers and therefore should not be run on any other machines.

Since the only file that is specific for a group of the servers is the `webserver.yaml` one, we only need to change it. To do so, let's open the `webserver.yaml` file and change content from `- hosts: all` to `- hosts: webserver`.

With only those three playbooks, we cannot proceed with creating our environment with three servers. Since we don't have a playbook to set up the database yet (we will see it in the next chapter), we will provision the two web servers (`ws01.fale.io` and `ws02.fale.io`) completely, and, for the database server, we will only provision the base system.

Before running the Ansible playbooks, we will need to provision the environment. To do so, create the following vagrant file:

```
Vagrant.configure("2") do |config|
  config.vm.define "ws01" do |ws01|
    ws01.vm.hostname = "ws01.fale.io"
  end
  config.vm.define "ws02" do |ws02|
```

```
    ws02.vm.hostname = "ws02.fale.io"
  end
  config.vm.define "db01" do |db01|
    db01.vm.hostname = "db01.fale.io"
  end
  config.vm.box = "centos/7"
end
```

Simply by running `vagrant up`, Vagrant will generate the whole environment for us. After a while of Vagrant outputting stuff in the shell, it should give you back the Command Prompt. When this happens, check that in the last few lines there were no errors, to be sure that everything went as expected.

Now that we have provisioned the environment, we can proceed by executing the `firstrun` playbook, which will ensure that our Ansible user is present and has the right SSH-key set up. To do so, we can run it with the following command:

```
$ ansible-playbook -i hosts firstrun.yaml
```

The following would be the result. Full output file is available on GitHub:

```
PLAY [all] *************************************************************

TASK [Gathering Facts] ************************************************
ok: [ws01.fale.io]
ok: [ws02.fale.io]
ok: [db01.fale.io]

TASK [Ensure ansible user exists] ***********************************
changed: [ws02.fale.io]
changed: [db01.fale.io]
changed: [ws01.fale.io]
...
```

As you can see, the output is very similar to what we received with a single host, but with one line per host at each step. In this case, all the machines were in the same state and the same steps have been performed, so we see that they all acted the same, but with more complex scenarios, you can have different machines returning different states on the same step. We can also execute the other two playbooks with similar results.

Regular expressions in the inventory file

When you have a large number of servers, it is common and helpful to give them predictable names, for instance, call all web servers wsXY or webXY, or call the database servers dbXY. If you do so, you can reduce the number of lines in your hosts file, increasing its readability. For instance, our hosts file can be simplified as follows:

```
[webserver]
ws[01:02].fale.io

[database]
db01.fale.io
```

In this example, we have used [01:02] that will match for all occurrences between the first number (01 in our case) and the last (02 in our case). In our case, the gain is not huge, but if you have 40 web servers, you can cut 39 lines from your hosts file.

In this section, we have seen how to create an inventory file, how to add groups to an Ansible inventory, how to leverage ranges to speed up the process of the creation of inventories, and how to run an Ansible playbook against an inventory. We will now see how to set variables in inventories and how to use them in our playbooks.

Working with variables

Ansible allows you to define variables in many ways, from a variable file within a playbook, by passing it from the Ansible command using the -e / --extra-vars option. You can also do it by passing it to an inventory file. You can define variables in an inventory file either on a per-host basis, for an entire group, or by creating a variable file in the directory where your inventory file exists.

Host variables

It's possible to declare variables for a specific host, declaring them in the hosts file. For instance, we may want to specify different engines for our web servers. Let's suppose that one needs to reply to a specific domain, while the other needs to reply to a different domain name. In this case, we would do it with the following hosts file:

```
[webserver]
ws01.fale.io domainname=example1.fale.io
ws02.fale.io domainname=example2.fale.io
```

```
[database]
db01.fale.io
```

Every time we execute a playbook with this inventory, Ansible will first read the inventory file, and it will assign, on a per-host basis, the value of the domainname variable. In this way, all playbooks running on web servers will be able to refer to the domainname variable.

Group variables

There are other cases where you want to set a variable that is valid for the whole group. Let's suppose that we want to declare the variable https_enabled to True and its value has to be equal for all web servers. In this case, we can create a [webserver:vars] section, so we will use the following hosts file:

```
[webserver]
ws01.fale.io
ws02.fale.io

[webserver:vars]
https_enabled=True

[database]
db01.fale.io
```

 Remember that host variables will override group variables if the same variable is declared in both spaces.

Variable files

Sometimes, you have a lot of variables to declare for each host and group, and the hosts file gets hard to read. In those cases, you can move the variables to specific files. For host-level variables, you'll need to create a file named the same as your host in the host_vars folder, while for group variables you'll have to use the group name for the file name and place them in the group_vars folder.

So, if we want to replicate the previous example of host-based variables using files, we will need to create the host_vars/ws01.fale.io file with the following content:

domainname=example1.fale.io

Then we create the `host_vars/ws02.fale.io` file with the following content:

```
domainname=example2.fale.io
```

While if we want to replicate the group-based variables example, we will need to have the `group_vars/webserver` file with the following content:

```
https_enabled=True
```

> Inventory variables follow a hierarchy; at the top of this is the common variable file (we discussed this in the previous section, *Working with inventory files*) that will override any of the host variables, group variables, and inventory variable files. After this come the host variables, which will override group variables; lastly, group variables will override inventory variable files.

Overriding configuration parameters with an inventory file

You can override some of Ansible's configuration parameters directly through the inventory file. These configuration parameters will override all the other parameters that are set either through `ansible.cfg`, environment variables, or set in the playbooks themselves. Variables passed to the `ansible-playbook/ansible` command have priority over any other variable, including the ones set in the inventory file.

The following is the list of some parameters you can override from an inventory file:

- `ansible_user`: This parameter is used to override the user that is used for communicating with the remote host. Sometimes, a certain machine needs a different user; in those cases, this variable will help you. For instance, if you are running Ansible from the *ansible* user but on the remote machine you need to connect to the *automation* user, setting `ansible_user=automation` will make it happen.
- `ansible_port`: This parameter will override the default SSH port with the user-specified port. Sometimes, sysadmin chooses to run SSH on a non-standard port. In this case, you'll need to instruct Ansible about the change. If in your environment the SSH port is 22022 instead of 22, you will need to use `ansible_port=22022`.

- `ansible_host`: This parameter is used to override the host for an alias. If you want to connect by the DNS name (that is: `ws01.fale.io`) to the `10.0.0.3` machine, but for some reason the DNS would not resolve the host properly, you can force Ansible to use this IP instead of what the DNS would resolve, by setting the `ansible_host=10.0.0.3` variable.
- `ansible_connection`: This specifies the connection type to the remote host. The values are SSH, Paramiko, or local. Even though Ansible could connect to the local machine using its SSH daemon, this wastes lots of resources. In those cases, you can specify `ansible_connection=local` so that Ansible will open a standard shell instead of SSH.
- `ansible_private_key_file`: This parameter will override the private key used for SSH; this will be useful if you want to use specific keys for a specific host. A common use case is if you have hosts spread across multiple data centers, multiple AWS regions, or different kinds of applications. Private keys can potentially be different in such scenarios.
- `ansible__type`: By default, Ansible uses the `sh` shell; you can override this using the `ansible_shell_type` parameter. Changing this to `csh`, `ksh`, and so on will make Ansible use the commands of that shell. This can be useful if you need to execute some `csh` or `ksh` scripts you have and it would be too costly to deal with them right away.

Working with dynamic inventories

There are environments where you have a system that creates and destroys machines automatically. We will see how to do this with Ansible in Chapter 5, *Going Cloud*. In such environments, the list of machines changes very quickly and keeping the hosts file becomes complicated. In this case, we can use dynamic inventories to solve the problem.

The idea behind dynamic inventories is that Ansible will not read the hosts file but instead execute a script that will return the list of hosts to Ansible in JSON format. This allows you, for instance, to query your cloud provider and ask it directly about the machines in your entire infrastructure that are running at any given moment.

Many scripts for the most common cloud providers are already available through Ansible at `https://github.com/ansible/ansible/tree/devel/contrib/inventory`, but you can create a custom script if you have different needs. The Ansible inventory scripts can be written in any language but, for consistency reasons, dynamic inventory scripts should be written in Python. Remember that these scripts need to be executable directly, so please remember to set them with the executable flag (`chmod + x inventory.py`).

Next, we will take a look at Amazon Web Services and DigitalOcean scripts that can be downloaded from the official Ansible repository.

Amazon Web Services

To allow Ansible to gather data from **Amazon Web Services (AWS)** about your EC2 instances, you need to download the following two files from Ansible's GitHub repository at `https://github.com/ansible/ansible`:

- The `ec2.py` inventory script
- The `ec2.ini` file, which contains the configuration for your EC2 inventory script

Ansible uses **Boto**, the AWS Python SDK, to communicate with AWS using APIs. To allow this communication, you need to export the `AWS_ACCESS_KEY_ID` and `AWS_SECRET_ACCESS_KEY` variables.

You can use the inventory in two ways:

- Pass it directly to an `ansible-playbook` command using the `-i` option and copy the `ec2.ini` file to your current directory where you are running the Ansible commands.
- Copy the `ec2.py` file to `/etc/ansible/hosts`, and make it executable using `chmod +x`, and copy the `ec2.ini` file to `/etc/ansible/ec2.ini`.

The `ec2.py` file will create multiple groups based on the region, availability zone, tags, and so on. You can check the contents of the inventory file by running `./ec2.py --list`.

Let's see an example playbook with the EC2 dynamic inventory, which will simply ping all machines in my account:

```
ansible -i ec2.py all -m ping
```

Since we executed the ping module, we expect the machines available in the configured account to reply to us. Since I have just one EC2 machine with the IP address 52.28.138.231 currently in my account, we can expect it to reply, and in fact the EC2 I have on my account responded with the following:

```
52.28.138.231 | SUCCESS => {
    "changed": false,
    "ping": "pong"
}
```

In the preceding example, we're using the `ec2.py` script instead of a static inventory file with the `-i` option and the ping command.

Similarly, you can use these inventory scripts to perform various types of operations. For example, you can integrate them with your deployment script to figure out all the nodes in a single zone and deploy to them if you're performing your deployment zone-wise (a zone represents a data center) in AWS.

If you simply want to know what the web servers in the cloud are and you've tagged them using a certain convention, you can do that by using the dynamic inventory script by filtering out the tags. Furthermore, if you have special scenarios that are not covered by your present script, you can enhance it to provide the required set of nodes in JSON format and then act on those nodes from the playbooks. If you're using a database to manage your inventory, your inventory script can query the database and dump a JSON. It could even sync with your cloud and update your database on a regular basis.

DigitalOcean

As we used the EC2 files in `https://github.com/ansible/ansible/tree/devel/contrib/inventory` to pull data from AWS, we can do the same for DigitalOcean. The only difference will be that we have to fetch the `digital_ocean.ini` and the `digital_ocean.py` files.

As before, we will need to tweak the `digital_ocean.ini` options, if needed, and to make the Python file executable. The only option that you'll probably need to change is `api_token`.

Now we can try to ping the two machines I've provisioned on DigitalOcean with the following:

```
ansible -i digital_ocean.py all -m ping
```

As expected, the two droplets I have on my account respond with the following:

```
188.166.150.79 | SUCCESS => {
    "changed": false,
    "ping": "pong"
}
46.101.77.55 | SUCCESS => {
    "changed": false,
    "ping": "pong"
}
```

We have now seen how easy it is to retrieve data from many different cloud providers.

Working with iterates in Ansible

You may have noticed that up to now we have never used cycles, so every time we had to do multiple, similar operations, we wrote the code multiple times. An example of this is the `webserver.yaml` code.

In fact, this is the last part of the `webserver.yaml` file:

```
  - name: Ensure HTTP can pass the firewall
    firewalld:
      service: http
      state: enabled
      permanent: True
      immediate: True
    become: True
  - name: Ensure HTTPS can pass the firewall
    firewalld:
      service: https
      state: enabled
      permanent: True
      immediate: True
    become: True
```

As you can see, the last two blocks of the `webserver.yaml` code do very similar operations: ensure that a certain port of the firewall is open.

Using standard iteration – with_items

Repeating code is not a problem per se, but it does not scale.

Ansible allows us to use iteration to improve the code both as clarity and maintainability.

To improve the preceding code, we can use a simple iteration: `with_items`.

This allows us to iterate in a list of items. At every iteration, the designated item of the list will be available to us in the item variable. This allows us to perform multiple similar operations in a single block.

We can therefore change the last part of the `webserver.yaml` code to the following:

```
- name: Ensure HTTP and HTTPS can pass the firewall
  firewalld:
    service: '{{ item }}'
    state: enabled
    permanent: True
    immediate: True
  become: True
  with_items:
    - http
    - https
```

We can execute it as follows:

```
ansible-playbook -i hosts webserver.yaml
```

We receive the following:

```
PLAY [all] ********************************************************

TASK [Gathering Facts] *******************************************
ok: [ws01.fale.io]
ok: [ws02.fale.io]

TASK [Ensure the HTTPd package is installed] *********************
ok: [ws02.fale.io]
ok: [ws01.fale.io]

TASK [Ensure the HTTPd service is enabled and running] ***********
ok: [ws01.fale.io]
ok: [ws02.fale.io]

TASK [Ensure HTTP and HTTPS can pass the firewall] **************
ok: [ws01.fale.io] (item=http)
ok: [ws02.fale.io] (item=http)
ok: [ws01.fale.io] (item=https)
ok: [ws02.fale.io] (item=https)

PLAY RECAP *******************************************************
ws01.fale.io               : ok=5 changed=0 unreachable=0 failed=0
ws02.fale.io               : ok=5 changed=0 unreachable=0 failed=0
```

As you can see, the output is slightly different than the previous execution. In fact, on the lines for operations with loops, we can see the `item` that was processed in that specific iteration of the `Ensure HTTP and HTTPS can pass the firewall` block.

We have now seen that we can iterate on a list of items, but Ansible allows us other kinds of iterations as well.

Using nested loops – with_nested

There are cases where you have to iterate all elements of a list with all items from other lists (Cartesian product). One case that is very common is when you have to create multiple folders in multiple paths. In our example, we will create the folders `mail` and `public_html` in the home folders of the users `alice` and `bob`.

We can do so with the following code snippet from the `with_nested.yaml` file; the full code is available on GitHub:

```
- hosts: all
  remote_user: ansible
  vars:
    users:
      - alice
      - bob
    folders:
      - mail
      - public_html
  tasks:
    - name: Ensure the users exist
      user:
        name: '{{ item }}'
      become: True
      with_items:
        - '{{ users }}'
    ...
```

Run this with the following:

```
ansible-playbook -i hosts with_nested.yaml
```

We receive the following result. Full output file is available on GitHub:

```
PLAY [all] ********************************************************

TASK [Gathering Facts] *******************************************
ok: [db01.fale.io]
ok: [ws02.fale.io]
ok: [ws01.fale.io]

TASK [Ensure the users exist] ***********************************
changed: [db01.fale.io] => (item=alice)
```

```
changed: [ws02.fale.io] => (item=alice)
changed: [ws01.fale.io] => (item=alice)
changed: [db01.fale.io] => (item=bob)
changed: [ws02.fale.io] => (item=bob)
changed: [ws01.fale.io] => (item=bob)
...
```

As you can see from the output, Ansible created the users alice and bob on all target machines, and it also created the folders $HOME/mail and $HOME/public_html for both users on all machines.

Fileglobs loop – with_fileglobs

Sometimes, we want to perform an action on every file present in a certain folder. This could be handy if you want to copy multiple files with similar names from one folder to another. To do so, you can create a file called with_fileglobs.yaml with the following code:

```
---
- hosts: all
  remote_user: ansible
  tasks:
    - name: Ensure the folder /tmp/iproute2 is present
      file:
        dest: '/tmp/iproute2'
        state: directory
      become: True
    - name: Copy files that start with rt to the tmp folder
      copy:
        src: '{{ item }}'
        dest: '/tmp/iproute2'
        remote_src: True
      become: True
      with_fileglob:
        - '/etc/iproute2/rt_*'
```

We can execute it with the following:

```
ansible-playbook -i hosts with_fileglobs.yaml
```

This results in the following output. Full output file is available on GitHub.

```
PLAY [all] *************************************************

TASK [Gathering Facts] ************************************
ok: [db01.fale.io]
```

```
ok: [ws02.fale.io]
ok: [ws01.fale.io]

TASK [Ensure the folder /tmp/iproute2 is present] **************
changed: [ws02.fale.io]
changed: [ws01.fale.io]
changed: [db01.fale.io]

TASK [Copy files that start with rt to the tmp folder] *********
changed: [ws01.fale.io] => (item=/etc/iproute2/rt_realms)
changed: [db01.fale.io] => (item=/etc/iproute2/rt_realms)
changed: [ws02.fale.io] => (item=/etc/iproute2/rt_realms)
changed: [ws01.fale.io] => (item=/etc/iproute2/rt_protos)
...
```

As for our goal, we have created the /tmp/iproute2 folder and populated with a copy of the files in the /etc/iproute2 folder. This pattern is often used to create backups of configurations.

Using an integer loop – with_sequence

Many times, you'll need to iterate over the integer numbers. An example could be to create ten folders called fileXY, where X and Y are sequential numbers from 1 to 10. To do so, we can create a file called with_sequence.yaml with the following code in it:

```
---
- hosts: all
  remote_user: ansible
  tasks:
  - name: Create the folders /tmp/dirXY with XY from 1 to 10
    file:
      dest: '/tmp/dir{{ item }}'
      state: directory
    with_sequence: start=1 end=10
    become: True
```

 Unlike the majority of Ansible commands, where we can use the single-line notation for objects and the standard YAML multi-line notation, with_sequence only supports the single line notation.

We can then execute it with the following:

```
ansible-playbook -i hosts with_sequence.yaml
```

We will receive this output:

```
PLAY [all] ********************************************************

TASK [Gathering Facts] ********************************************
ok: [ws02.fale.io]
ok: [ws01.fale.io]
ok: [db01.fale.io]

TASK [Create the folders /tmp/dirXY with XY from 1 to 10] ******
changed: [ws01.fale.io] => (item=1)
changed: [db01.fale.io] => (item=1)
changed: [ws02.fale.io] => (item=1)
changed: [ws01.fale.io] => (item=2)
changed: [db01.fale.io] => (item=2)
changed: [ws02.fale.io] => (item=2)
changed: [ws01.fale.io] => (item=3)
...
```

Ansible supports many more types of loop, but since they are used far less, you can refer directly to the official documentation about loops
at: http://docs.ansible.com/ansible/playbooks_loops.html.

Summary

In this chapter, we have explored a large number of concepts that will help scale your infrastructure beyond the single node. We started with inventory files used to instruct Ansible about our machines, and then we covered how to have host-specific and group-specific variables while running the same command on multiple heterogeneous hosts. We then moved on to dynamic inventories that are populated directly by some other system (usually a cloud provider). Finally, we analyzed multiple kinds of iteration in the Ansible playbooks.

In the next chapter, we will structure our Ansible files in a saner way to ensure maximum readability. To do this, we introduce roles that simplify the management of complex environments even more.

Handling Complex Deployment

4

So far, we've seen how to write basic Ansible playbooks, options associated with playbooks, practices to develop playbooks using Vagrant, and how to test playbooks at the end of the process. We've now got a framework for you and your team to learn and start developing Ansible playbooks. Consider this a similar process to learning how to drive a car from your driving-school instructor. You start by learning how to control the car with the steering wheel, then you slowly begin to control the brakes, and finally, you start maneuvering the gears, and hence, the speed of your car. Once you've done this over a period of time, with more and more practice on different kinds of roads (such as flat, hilly, muddy, potholed, and so on) and by driving different cars, you gain expertise, fluency, speed, and, basically, enjoy the overall experience. From this chapter onward, we will up the gear by digging deeper into Ansible and urge you to practice and try out more examples to get comfortable with it.

You must be wondering why this chapter is named the way it is. The reason for this is that, so far, we've not yet reached a stage where you can deploy playbooks in production, especially in complex situations. Complex situations include those where you have to interact with several hundred or thousands of machines where each group of machines is dependent on another group or groups of machines. These groups may be dependent on one another for all or some transactions, to perform secure complex data backups and replications with masters and slaves. In addition, there are several interesting and rather compelling features of Ansible that we've not yet looked at. In this chapter, we will cover all of them with examples. Our aim is that, by the end of this chapter, you should have a clear idea of how to write playbooks that can be deployed in production from a configuration management perspective. The following chapters will add to what we've learned, to enhance the experience of using Ansible.

The following topics will be covered in the chapter:

- Working with the `local_action` feature as well as other task delegation and conditionals strategies
- Working with include, handlers and roles
- Transforming your playbooks
- Jinja filters
- Security management tips and tools

Technical requirements

You can download all of the files from this book's GitHub repository at `https://github.com/PacktPublishing/Learning-Ansible-2.X-Third-Edition/tree/master/Chapter04`.

Working with the local_action feature

The `local_action` feature of Ansible is a powerful one, especially when we think of **orchestration**. This feature allows you to run certain tasks locally on the machine that runs Ansible.

Consider the following situations:

- Spawning a new machine or creating a JIRA ticket
- Managing your command center(s) in terms of installing packages and setting up configurations
- Calling a load balancer API to disable a certain web server entry from the load balancer

These are tasks that can, usually, be run on the same machine that runs the `ansible-playbook` command rather than logging in to a remote box and running these commands.

Let's look at an example. Suppose you want to run a shell module on your local system where you are running your Ansible playbook. The `local_action` option comes into the picture in such situations. If you pass the module name and the module argument to `local_action`, it will run that module locally.

Let's see how this option works with the `shell` module. Consider the following code that shows the output of the `local_action` option:

```
---
- hosts: database
  remote_user: vagrant
  tasks:
    - name: Count processes running on the remote system
      shell: ps | wc -l
      register: remote_processes_number
    - name: Print remote running processes
      debug:
        msg: '{{ remote_processes_number.stdout }}'
    - name: Count processes running on the local system
      local_action: shell ps | wc -l
      register: local_processes_number
    - name: Print local running processes
      debug:
        msg: '{{ local_processes_number.stdout }}'
```

We can now save it as `local_action.yaml` and run it with the following:

```
ansible-playbook -i hosts local_action.yaml
```

We receive the following result:

```
PLAY [database] ***************************************************
TASK [Gathering Facts] *******************************************
ok: [db01.fale.io]
TASK [Count processes running on the remote system] ************
changed: [db01.fale.io]
TASK [Print remote running processes] **************************
ok: [db01.fale.io] => {
    "msg": "6"
}
TASK [Count processes running on the local system] ************
changed: [db01.fale.io -> localhost]
TASK [Print local running processes] **************************
ok: [db01.fale.io] => {
    "msg": "9"
}
PLAY RECAP ****************************************************
db01.fale.io                : ok=5 changed=2 unreachable=0 failed=0
```

As you can see, the two commands provided us with different numbers, since they have been executed on different hosts. You can run any module with `local_action`, and Ansible will make sure that the module is run locally on the box where the `ansible-playbook` command is run. Another simple example you can (and should!) try is running two tasks:

- `uname` on the remote machine (`db01` in the preceding case)
- `uname` on the local machine but with `local_action` enabled

This will crystalize the idea of `local_action` further.

Ansible provides another method to delegate certain actions to a specific (or different) machine: the `delegate_to` system.

Delegating a task

Sometimes, you will want to execute an action on a different system. This could be, for instance, a database node while you are deploying something on an application server node or to the localhost. To do so, you can just add the `delegate_to: HOST` property to your task and it will be run on the proper node. Let's rework the previous example to achieve this:

```yaml
---
- hosts: database
  remote_user: vagrant
  tasks:
    - name: Count processes running on the remote system
      shell: ps | wc -l
      register: remote_processes_number
    - name: Print remote running processes
      debug:
        msg: '{{ remote_processes_number.stdout }}'
    - name: Count processes running on the local system
      shell: ps | wc -l
      delegate_to: localhost
      register: local_processes_number
    - name: Print local running processes
      debug:
        msg: '{{ local_processes_number.stdout }}'
```

Saving it as `delegate_to.yaml`, we can run it with the following:

```
ansible-playbook -i hosts delegate_to.yaml
```

We will receive the same output as we did in the previous example:

```
PLAY [database] ************************************************

TASK [Gathering Facts] *****************************************
ok: [db01.fale.io]
TASK [Count processes running on the remote system] ***********
changed: [db01.fale.io]
TASK [Print remote running processes] *************************
ok: [db01.fale.io] => {
    "msg": "6"
}

TASK [Count processes running on the local system] ***********
changed: [db01.fale.io -> localhost]

TASK [Print local running processes] **************************
ok: [db01.fale.io] => {
    "msg": "9"
}
PLAY RECAP ***************************************************
db01.fale.io              : ok=5 changed=2 unreachable=0 failed=0
```

In this example we have seen how we can perform actions on the remote host and the local host in the same playbook. This becomes handy in complex procedures where some parts of the procedure need to be executed by the local machine or any other machine you can connect to.

Working with conditionals

Until now, we have only seen how playbooks work and how tasks are executed. We have also seen that Ansible executes all of these tasks sequentially. However, this would not help you while writing an advanced playbook that contains tens of tasks and have to execute only a subset of these tasks. For example, let's say you have a playbook that will install Apache HTTPd server on the remote host. Now, the Apache HTTPd server has a different package name for a Debian-based operating system, and it's called `apache2`; for a Red Hat-based operating system, it's called `httpd`.

Having two tasks, one for the `httpd` package (for Red Hat-based systems) and the other for the `apache2` package (for Debian-based systems) in a playbook will make Ansible install both packages, and this execution will fail, as `apache2` will not be available if you're installing on a Red Hat-based operating system. To overcome such problems, Ansible provides conditional statements that help to run a task only when a specified condition is met. In this case, we do something similar to the following pseudocode:

```
If os = "redhat"
   Install httpd
Else if os = "debian"
   Install apache2
End
```

While installing `httpd` on a Red Hat-based operating system, we first check whether the remote system is running a Red Hat-based operating system, and if it is, we then install the `httpd` package; otherwise, we skip the task. Without wasting your time, let's dive into an example playbook called `conditional_httpd.yaml` with the following content:

```
---
- hosts: webserver
  remote_user: vagrant
  tasks:
    - name: Print the ansible_os_family value
      debug:
        msg: '{{ ansible_os_family }}'
    - name: Ensure the httpd package is updated
      yum:
        name: httpd
        state: latest
      become: True
      when: ansible_os_family == 'RedHat'
    - name: Ensure the apache2 package is updated
      apt:
        name: apache2
        state: latest
      become: True
      when: ansible_os_family == 'Debian'
```

Run it with the following:

```
ansible-playbook -i hosts conditional_httpd.yaml
```

This is the result. Full code output file is available on GitHub:

```
PLAY [webserver] *********************************************

TASK [Gathering Facts] ***************************************
ok: [ws03.fale.io]
ok: [ws02.fale.io]
ok: [ws01.fale.io]

TASK [Print the ansible_os_family value] *********************
ok: [ws01.fale.io] => {
  "msg": "RedHat"
}
ok: [ws02.fale.io] => {
  "msg": "RedHat"
}
ok: [ws03.fale.io] => {
  "msg": "Debian"
}
...
```

As you can see, I've created a new server (ws03) for this example that is Debian-based. As expected, the installation of the httpd package was performed on the two CentOS nodes, while the installation of the apache2 package was performed on the Debian node.

> Ansible only distinguishes between a few families (AIX, Alpine, Altlinux, Archlinux, Darwin, Debian, FreeBSD, Gentoo, HP-UX, Mandrake, Red Hat, Slackware, Solaris, and Suse at the time of writing this book); for this reason, a CentOS machine has an ansible_os_family value: RedHat.

Likewise, you can match for different conditions as well. Ansible supports equal to (==), different than (!=), greater than (>), smaller than (<), greater than or equal to (>=), and smaller than or equal to (<=).

The operators we have seen so far will match the entire content of the variable, but what if you just want to check whether a particular character or a string is present in a variable? To perform these kinds of checks, Ansible provides the in and not operators. You can also match multiple conditions using the and and or operators. The and operator will make sure that all conditions are matched before executing this task, whereas the or operator will make sure that there is a match for at least one of the conditions.

Boolean conditionals

Apart from string matching, you can also check whether a variable is True. This type of validation will be useful when you want to check whether a variable was assigned a value. You can even execute a task based on the Boolean value of a variable.

For example, let's put the following code in a file called `crontab_backup.yaml`:

```
---
- hosts: all
  remote_user: vagrant
  vars:
    backup: True
  tasks:
    - name: Copy the crontab in tmp if the backup variable is true
      copy:
        src: /etc/crontab
        dest: /tmp/crontab
        remote_src: True
      when: backup
```

We can execute it with the following:

```
ansible-playbook -i hosts crontab_backup.yaml
```

And we obtain the following:

```
PLAY [all] ****************************************************

TASK [Gathering Facts] ***************************************
ok: [ws03.fale.io]
ok: [ws01.fale.io]
ok: [db01.fale.io]
ok: [ws02.fale.io]

TASK [Copy the crontab in tmp if the backup variable is true]
changed: [ws03.fale.io]
changed: [ws02.fale.io]
changed: [ws01.fale.io]
changed: [db01.fale.io]

PLAY RECAP ***************************************************
db01.fale.io        : ok=2 changed=1 unreachable=0 failed=0
ws01.fale.io        : ok=2 changed=1 unreachable=0 failed=0
ws02.fale.io        : ok=2 changed=1 unreachable=0 failed=0
ws03.fale.io        : ok=2 changed=1 unreachable=0 failed=0
```

We can change the command, slightly, to this:

```
ansible-playbook -i hosts crontab_backup.yaml --extra-vars="backup=False"
```

And we will receive this output:

```
PLAY [all] ***************************************************

TASK [Gathering Facts] **************************************
ok: [ws03.fale.io]
ok: [ws01.fale.io]
ok: [db01.fale.io]
ok: [ws02.fale.io]

TASK [Copy the crontab in tmp if the backup variable is true]
skipping: [ws01.fale.io]
skipping: [ws02.fale.io]
skipping: [ws03.fale.io]
skipping: [db01.fale.io]

PLAY RECAP ***************************************************
db01.fale.io              : ok=1  changed=0  unreachable=0  failed=0
ws01.fale.io              : ok=1  changed=0  unreachable=0  failed=0
ws02.fale.io              : ok=1  changed=0  unreachable=0  failed=0
ws03.fale.io              : ok=1  changed=0  unreachable=0  failed=0
```

As you can see, in the first case, the operation was executed, while in the second case, it was skipped. We could have overwritten the back-up value using a configuration file, a host variable, or a group variable.

> If checked in this way and if the variable is not set, Ansible will assume it to be False.

Checking whether a variable is set

Sometimes, you find yourself having to use a variable in a command. Every time you do so, you have to ensure that the variable is *set*. This is because some commands could be catastrophic if called with an *unset* variable (that is, if you execute rm -rf $VAR/* and $VAR is not set or empty, it will nuke your machine). To do so, Ansible provides a way to check whether a variable is defined.

We could improve the previous example in the following way:

```
---
- hosts: all
  remote_user: ansible
  vars:
    backup: True
  tasks:
    - name: Check if the backup_folder is set
      fail:
        msg: 'The backup_folder needs to be set'
      when: backup_folder is not defined or backup_folder == ""
    - name: Copy the crontab in tmp if the backup variable is true
      copy:
        src: /etc/crontab
        dest: '{{ backup_folder }}/crontab'
        remote_src: True
      when: backup
```

As you can see, we have used the fail module that allows us to put the Ansible playbook in a failure state if the `backup_folder` variable is not set.

Working with include

The `include` feature helps you to reduce duplicity while writing tasks. This also allows us to have smaller playbooks by including reusable code in separate tasks using the **Don't Repeat Yourself** (**DRY**) principle.

To trigger the inclusion of another file, you need to put the following under the tasks object:

```
- include: FILENAME.yaml
```

You can also pass some variables to the included file. To do so, we can specify them in the following way:

```
- include: FILENAME.yaml variable1="value1" variable2="value2"
```

Keeping your variable clean and minimalistic using the include statement, and by observing the DRY principle will allow you to write Ansible code that is going to be far more maintainable and easy to follow.

Working with handlers

In many situations, you will have a task or a group of tasks that change certain resources on the remote machines that need to trigger an event to become effective. For example, when you change a service configuration, you will need to restart or reload the service itself. In Ansible, you can trigger this event using the `notify` action.

Every handler task will run at the end of the playbook if notified. For example, you changed your HTTPd server configuration multiple times and you want to restart the HTTPd service so that the changes are applied. Now, restarting HTTPd every single time you make a configuration change is not a good practice; it is not a good practice to restart the server even if no changes have been made to its configurations. To deal with such a situation, you can notify Ansible to restart the HTTPd service on every configuration change, but Ansible will make sure that, no matter how many times you notify it for the HTTPd restart, it will call that task just once after all other tasks complete. Let's change the `webserver.yaml` file we created in the previous chapters a little bit in the following way; the full code is available on GitHub:

```
---
- hosts: webserver
  remote_user: vagrant
  tasks:
    - name: Ensure the HTTPd package is installed
      yum:
        name: httpd
        state: present
      become: True
    - name: Ensure the HTTPd service is enabled and running
      service:
        name: httpd
        state: started
        enabled: True
      become: True
    - name: Ensure HTTP can pass the firewall
      firewalld:
  service: http
        state: enabled
        permanent: True
        immediate: True
      become: True
  ...
```

Run this script with the following:

```
ansible-playbook -i hosts webserver.yaml
```

We will have the following output. Full code output file is available on GitHub:

```
PLAY [webserver] ****************************************

TASK [Gathering Facts] **********************************
ok: [ws01.fale.io]
ok: [ws02.fale.io]

TASK [Ensure the HTTPd package is installed] ****************
ok: [ws01.fale.io]
ok: [ws02.fale.io]

TASK [Ensure the HTTPd service is enabled and running] *******
changed: [ws02.fale.io]
changed: [ws01.fale.io]

...
```

In this case, the handler has been triggered from the configuration file change. But if we run it a second time, the configuration will not change and therefore, we will have the following result:

```
PLAY [webserver] ****************************************

TASK [Gathering Facts] **********************************
ok: [ws01.fale.io]
ok: [ws02.fale.io]

TASK [Ensure the HTTPd package is installed] ****************
ok: [ws01.fale.io]
ok: [ws02.fale.io]

TASK [Ensure the HTTPd service is enabled and running] *******
ok: [ws02.fale.io]
ok: [ws01.fale.io]

TASK [Ensure HTTP can pass the firewall] ********************
ok: [ws02.fale.io]
ok: [ws01.fale.io]

TASK [Ensure HTTPd configuration is updated] ****************
ok: [ws02.fale.io]
ok: [ws01.fale.io]

PLAY RECAP **********************************************
ws01.fale.io            : ok=5 changed=0 unreachable=0 failed=0
ws02.fale.io            : ok=5 changed=0 unreachable=0 failed=0
```

As you can see, this time no handler has been executed, since all steps that could have triggered their execution returned without changes, so the handlers were not needed. Remember this behavior to ensure you are not surprised by handlers not executed.

Working with roles

We have seen how we can automate simple tasks, but what we have seen up till now will not solve all of your problems. This is because playbooks are very good at executing operations, but they are not very good at configuring huge amounts of machines, because they will soon become messy. To solve this, Ansible has **roles**.

My definition of a role is a set of playbooks, templates, files, or variables used to achieve a specific goal. For instance, we could have a database role and a web server role so that those configurations stay cleanly separated.

Before starting to look inside a role, let's talk about organizing a project.

Organizing a project

In the last few years, I've worked on multiple Ansible repositories for multiple organizations, and many of them were very chaotic. To ensure that your repository is easy to manage, I'm going to give you a template that I always use.

First of all, I always create three files in the `root` folder:

- `ansible.cfg`: A small configuration file to explain to Ansible where to find the files in our folder structure
- `hosts`: The hosts file we have already seen in the previous chapters
- `master.yaml`: A playbook that aligns the whole infrastructure

In addition to those three files, I create two folders:

- `playbooks`: This will contain the playbooks and a folder called `groups` for groups management.
- `roles`: This will contain all of the roles we need.

To clarify this, let's use the Linux `tree` command to see the structure of an Ansible repository for a simple web application needing web servers and database servers:

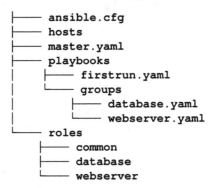

```
├──── ansible.cfg
├──── hosts
├──── master.yaml
├──── playbooks
│     ├──── firstrun.yaml
│     └──── groups
│           ├──── database.yaml
│           └──── webserver.yaml
└──── roles
      ├──── common
      ├──── database
      └──── webserver
```

As you can see, I've added a `common` role as well. This is very useful for putting in all of the things that should be performed for every server. Usually, I configure NTP, `motd`, and other similar services in this role, as well as the machine hostname.

We will now see how to structure a role.

Anatomy of a role

The structure of folders in a role is standard and you cannot change it much.

The most important folder within the role is the `tasks` folder, because this is the only mandatory folder in it. It has to contain a `main.yaml` file that will be the list of tasks to be executed. Other folders that are often present in the roles are templates and files. The first one will be used to store templates used by the **template task**, while the second will be used to store files that are used by the **copy task**.

Transforming your playbooks in a full Ansible project

Let's see how to transform the three playbooks we used to set up our web infrastructure (`common_tasks.yaml`, `firstrun.yaml`, and `webserver.yaml`) to fit this file organization. We have to remember that we also used two files (`index.html.j2` and `motd`) in those roles, so we have to place these files properly too.

First, we are going to create the folder structure we have seen in the previous paragraph.

The easiest playbook to port is `firstrun.yaml`, since we only need to copy it into the `playbooks` folder. This playbook will remain a playbook because it's a set of operations that will have to be run just one time for each server.

We now move to the `common_tasks.yaml` playbook, which will need a little bit of a rework to match the role paradigm.

Transforming a playbook into a role

The first thing we need to do is to create the `roles/common/tasks` and `roles/common/templates` folders. In the first one, we will add the following `main.yaml` file. The full code is available on GitHub:

```
---
- name: Ensure EPEL is enabled
  yum:
    name: epel-release
    state: present
  become: True
- name: Ensure libselinux-python is present
  yum:
    name: libselinux-python
    state: present
  become: True
- name: Ensure libsemanage-python is present
  yum:
    name: libsemanage-python
    state: present
  become: True
...
```

As you can see, this is very similar to our `common_tasks.yaml` playbooks. In fact, there are only two differences:

- The `hosts`, `remote_user`, and `tasks` lines (lines 2, 3, and 4) have been deleted.
- The indentation of the rest of the file has been fixed accordingly.

In this role, we used the template task to create a `motd` file on the server with the IP of the machine and other interesting information. For this reason, we need to create `roles/common/templates` and put the `motd` template in it.

At this point, our common task will have this structure:

```
common/
├───── tasks
│      └────── main.yaml
└────── templates
        └────── motd
```

We now need to instruct Ansible on the machines that will need to perform all of the tasks specified in the `common` role. To do so, we should look at the playbooks/groups directory. In this directory, it is handy to have one file for each group of logically similar machines (that is, machines that are performing the same kind of operation), in our case, the database and the web server.

So, let's create a `database.yaml` file in `playbooks/groups` with the following content:

```
---
- hosts: database
  user: vagrant
  roles:
  - common
```

Create a `webserver.yaml` file in the same folder with the following content:

```
---
- hosts: webserver
  user: vagrant
  roles:
  - common
```

As you can see, these files specify the group of hosts that we want to operate on, the remote user to use on those hosts, and the roles that we want to execute.

Helper files

When we created the `hosts` file in the previous chapter, we noticed that it helps to simplify our command lines. So, let's start copying the hosts files we previously used in the `root` folder of our Ansible repository. Up to now, we have always specified the path of this file on the command line. This is no longer necessary if we create an `ansible.cfg` file that tells Ansible the location of our `hosts` file. For this reason, let's create an `ansible.cfg` file in the root of our Ansible repository with the following content:

```
[defaults]
inventory = hosts
```

```
host_key_checking = False
roles_path = roles
```

In this file, we have also specified another two variables in addition to the `inventory` one that we already talk about, and those are `host_key_checking` and `roles_path`.

The `host_key_checking` flag is useful to not require the verification of the remote system SSH key. This is not suggested for use in production, since the usage of a public key propagation system is suggested for such environments, but it is very handy in testing environments, since it will help you to reduce the time Ansible hangs around waiting for user input.

`roles_path` is used to tell Ansible where to find the roles for our playbooks.

I usually add one additional file, which is `master.yaml`. I find it very useful as you will often need to keep your infrastructure aligned with your Ansible code. To do it in a single command, you'll need a file that will run all of the files in playbooks/groups. So, let's create a `master.yaml` file in the Ansible repository `root` folder with the following content:

```
---
- import_playbook: playbooks/groups/database.yaml
- import_playbook: playbooks/groups/webserver.yaml
```

At this point, we can execute the following:

```
ansible-playbook master.yaml
```

The result will be the following. Full code output file is available on GitHub:

```
PLAY [database] *********************************************

TASK [Gathering Facts] *************************************
ok: [db01.fale.io]

TASK [common : Ensure EPEL is enabled] ********************
ok: [db01.fale.io]

TASK [common : Ensure libselinux-python is present] *********
ok: [db01.fale.io]

TASK [common : Ensure libsemanage-python is present] ********
ok: [db01.fale.io]

TASK [common : Ensure we have last version of every package] *
ok: [db01.fale.io]
...
```

As you can see in the preceding output, the actions listed in the common role have been executed on the node in the database group first and then on the nodes in the webserver group.

Transforming the webserver role

As we transformed the common playbook into the common role, we can do the same for the webserver role.

In roles, we need to have the webserver folder with the tasks subfolder inside it. In this folder, we have to put the main.yaml file containing the tasks copied from the playbooks. The following is the code snippet; the full code is available on GitHub:

```
---
- name: Ensure the HTTPd package is installed
  yum:
    name: httpd
    state: present
  become: True
- name: Ensure the HTTPd service is enabled and running
  service:
    name: httpd
    state: started
    enabled: True
  become: True
- name: Ensure HTTP can pass the firewall
  firewalld:
    service: http
    state: enabled
    permanent: True
    immediate: True
  become: True
...
```

In this role, we have used multiple tasks that will need additional resources to work properly; more specifically, we need to do the following:

- Put the website.conf file in roles/webserver/files.
- Put the index.html.j2 template in roles/webserver/templates.
- Create the Restart HTTPd handler.

The first two should be pretty straightforward. The first one, in fact, is an empty file (we have not yet put anything in it since the default configuration was good enough for our use) and the `index.html.j2` file should contain the following content:

```
<html>
    <body>
        <h1>Hello World!</h1>
        <p>This page was created on {{ ansible_date_time.date }}.</p>
        <p>This machine can be reached on the following IP addresses</p>
        <ul>
{% for address in ansible_all_ipv4_addresses %}
            <li>{{ address }}</li>
{% endfor %}
        </ul>
    </body>
</html>
```

Handlers in roles

The last thing we need to do to complete this role is to create the handler for the `Restart HTTPd` notification. To do so, we will need to create a `main.yaml` file in `roles/webserver/handlers` with the following content:

```
---
- name: Restart HTTPd
  service:
    name: httpd
    state: restarted
  become: True
```

As you may have noticed, this is very similar to the handler we used in the playbook, if not for the file location and indentation.

The only thing that we still need to do to make our role applicable is to add the entry to the `playbooks/groups/webserver.yaml` file so that Ansible is informed that the servers in the web server group should apply the web server role as well as the common role. Our `playbooks/groups/webserver.yaml` file will need to be like the following:

```
---
- hosts: webserver
  user: ansible
  roles:
  - common
  - webserver
```

We could now execute `master.yaml` again to apply the web server role to the relevant servers, but we can also just execute `playbooks/groups/webserver.yaml`, since the change we just did is relevant only to this group of servers. To do so, we run the following:

```
ansible-playbook playbooks/groups/webserver.yaml
```

We should receive output similar to the following. Full code output file is available on GitHub:

```
PLAY [webserver] ********************************************

TASK [Gathering Facts] *************************************
ok: [ws01.fale.io]
ok: [ws02.fale.io]

TASK [common : Ensure EPEL is enabled] ********************
ok: [ws01.fale.io]
ok: [ws02.fale.io]

TASK [common : Ensure libselinux-python is present] **********
ok: [ws01.fale.io]
ok: [ws02.fale.io]

TASK [common : Ensure libsemanage-python is present] *********
ok: [ws01.fale.io]
ok: [ws02.fale.io]


. . .
```

As you can see in the preceding output, both the `common` and `webserver` roles have been applied to the `webserver` nodes.

It's very important to apply all roles concerning a specific node and not just the one you changed, because, more often than not, when there is a problem on one or more nodes in a group but not on other nodes of the same group, the problem is that some roles have been applied unequally in the group. Only applying all concerned roles to a group will grant you the equality of the nodes of that group.

Execution strategies

Before Ansible 2, every task needed to be executed (and completed) on each machine before Ansible issued a new task to all machines. This meant that, if you were performing tasks on a hundred machines and one of them was under-performing, all of the machines would run at the under-performing machine's speed.

With Ansible 2, execution strategies have been made modular and pluggable; therefore, you can now choose which execution strategy you prefer for your playbooks. You can also write custom execution strategies, but this is beyond the scope of this book. At the moment (in Ansible 2.7), there are only three execution strategies, **linear**, **serial**, and **free**:

- **Linear execution**: This strategy behaves exactly as Ansible did prior to version 2. This is the default strategy.
- **Serial execution**: This strategy will take a subset of hosts (the default is five) and execute all tasks against those hosts before moving to the next subset and starting from the beginning. This kind of execution strategy could help you to work on a limited number of hosts so that you always have some hosts that are available to your users. If you are looking for this kind of deployment, you will need a load balancer in front of your hosts that needs to be informed about which nodes are in maintenance at every given moment.
- **Free execution**: This strategy will serve a new task to each host as soon as that host has completed the previous task. This will allow faster hosts to complete the playbook before slower nodes. If you choose this execution strategy, you have to remember that some tasks could require a previous task to be completed on all nodes (for instance, clustering databases require all database nodes to have the database installed and running), and in this case, they will probably fail.

The Ansible template – Jinja filters

We have seen in `Chapter 2`, *Automating Simple Tasks*, this templates allow you to dynamically complete your playbook and place files on servers based on dynamic data such as `host` and `group` variables. In this section, we will move forward and see how **Jinja2 filters** work with Ansible.

Jinja2 filters are simple Python functions that take some arguments, process them, and return the result. For example, consider the following command:

```
{{ myvar | filter }}
```

In the preceding example, `myvar` is a variable; Ansible will pass `myvar` to the Jinja2 filter as an argument. The Jinja2 filter will then process it and return the resulting data. Jinja2 filters even accept additional arguments, as follows:

```
{{ myvar | filter(2) }}
```

In this example, Ansible will now pass two arguments, that is, `myvar` and `2`. Likewise, you can pass multiple arguments to filters separated by commas.

Ansible supports a wide variety of Jinja2 filters, and we will see some of the important Jinja2 filters that you might need to use while writing your playbook.

Formatting data using filters

Ansible supports Jinja2 filters to format data to JSON or YAML. You pass a dictionary variable to this filter, and it will format your data into JSON or YAML. For example, consider the following command line:

```
{{ users | to_nice_json }}
```

In the preceding example, `users` is the variable and `to_nice_json` is the Jinja2 filter. As we saw earlier, Ansible will internally pass `users` as an argument to the Jinja2 filter `to_nice_json`. Likewise, you can format your data into YAML as well, using the following command:

```
{{ users | to_nice_yaml }}
```

Defaulting undefined variables

We have seen in the previous sections that it is always wise to check whether a variable is defined before using it. We can set a `default` value for the variable so that, instead of failing, Ansible will use that value if the variable is not defined. To do so, we use this:

```
{{ backup_disk | default("/dev/sdf") }}
```

This filter will not assign the `default` value to the variable; it will only pass the `default` value to the current task where it is being used. Let's look at a few more examples of Jinja filters themselves before closing this section:

- Execute this to get a random character from a list:

```
{{ ['a', 'b', 'c', 'd'] | random }}
```

- Execute this to get a random number from `0` to `100`:

```
{{ 100 | random }}
```

- Execute this to get a random number from `10` to `50`:

  ```
  {{ 50   | random(10) }}
  ```

- Execute this to get a random number from `20` to `50` in steps of `10`:

  ```
  {{ 50 | random(20, 10) }}
  ```

- Concatenating a list to string using filters: Jinja2 filters allow you to concatenate a list to a string using the join filter. This filter takes a separator as an extra argument. If you do not specify a separator, then the filter will combine all elements of the list together without any separation. Consider the following example:

  ```
  {{ ["This", "is", "a", "string"] | join(" ") }}
  ```

The preceding filter will result in a `This is a string` output. You can specify any separator you want instead of a white space.

When it comes to encoding or decoding data using filters, you can encode or decode data using filters as follows:

- Encode your data to `base64` using the `b64encode` filter:

  ```
  {{ variable | b64encode }}
  ```

- Decode an encoded `base64` string using the `b64decode` filter:

  ```
  {{ "aGFoYWhhaGE=" | b64decode }}
  ```

Security management

The last section in this chapter is about **security management**. If you tell your system administrator that you want to introduce a new feature or a tool, one of the first questions they would ask you would be this: *"What security feature(s) are present with your tool?"* We'll try to answer this question from an Ansible perspective in this section. Let's look at them in greater detail.

Using Ansible Vault

Ansible Vault is an exciting feature of Ansible that was introduced in Ansible Version 1.5. This allows you to have encrypted passwords as part of your source code. A recommended practice is to *not* have passwords (as well as any other sensitive information such as private keys and SSL certificates) in plaintext as part of your repository, because anyone who checks out your repository can view your passwords. Ansible Vault can help you to secure your confidential information by encrypting and decrypting them on your behalf.

Ansible Vault supports an interactive mode in which it will ask you for the password or a non-interactive mode where you will have to specify the file containing the password and Ansible Vault will read it directly.

For these examples, we will use the password `ansible`, so let's start creating a hidden file called `.password` with the `ansible` string in it. To do so, let's execute the following:

```
echo 'ansible' > .password
```

We can now create `ansible-vault` both in the interactive and non-interactive modes. If we want to do it in interactive mode, we will need to execute the following:

```
ansible-vault create secret.yaml
```

Ansible will ask us for the vault password and then confirm it. Later, it will open the default text editor (in my case, **vi**) to add the content in clear. I have used the password `ansible` and the text is `This is a password protected file`. We can now save and close the editor and check that `ansible-vault` has encrypted our content; in fact, we can run this:

```
cat secret.yaml
```

It will output the following:

```
$ANSIBLE_VAULT;1.1;AES256
65396465353561366663565333333396238323734623462626563346135366434653261356639336593365
3263633761383434363766613962386637383465643130320a633862343137306563323236313393013930
326535333316238633731363333386463232373935353935352313366653538633538643737353939336593365
3433356539333232650a643737326362396333362342323365303036633665333034653437376437394373943739
63373438316435626138646462366436633136393033333306330130393761343531313238653733307330
6333663133353730303561303535353562306535333346364613830
```

In the same way, we can invoke the `ansible-vault` command with the `- vault-password-file=VAULT_PASSWORD_FILE` option to specify our `.password` file. We can, for instance, edit our `secret.yaml` file with this command:

```
ansible-vault --vault-password-file=.password edit secret.yaml
```

This will open your default text editor where you'll be able to change the file as if it were a plain file. When you save the file, Ansible Vault will perform the encryption before saving it, assuring the confidentiality of your content.

Sometimes, you need to look at the content of a file, but you don't want to open it in a text editor, so you usually use the `cat` command. Ansible Vault has a similar feature called `view`, so you can run this:

```
ansible-vault --vault-password-file=.password view secret.yaml
```

Ansible Vault allows you to decrypt a file, replacing its encrypted content with its plain text content. To do so, you can execute the following:

```
ansible-vault --vault-password-file=.password decrypt secret.yaml
```

At this point, we can the `cat` command on the `secret.yaml` file and the result is the following:

```
This is a password protected file
```

Ansible Vault also gives you the capability to encrypt files that already exist. This is particularly useful if you want to develop all of your files on a trusted machine (for instance, your own local machine) in a clear text to improve your efficiency and then encrypt all sensitive files afterword. To do so, you can execute this:

```
ansible-vault --vault-password-file=.password encrypt secret.yaml
```

You can now check that the `secret.yaml` file is encrypted again.

The last option of the Ansible Vault is very important since it's a `rekey` function. This function will allow you to change the encryption key in a single command. You could perform the same operation with two commands (decrypt the `secret.yaml` file with the **old key** and then encrypt it with the **new key**) but being able to perform it in a single step has major advantages, since the file in its cleartext form will not be stored on the disk at any moment of the process.

To do so, we need a file containing the new password (in our case, the file called `.newpassword` and containing the `ansible2` string), and you need to execute the following command:

```
ansible-vault --vault-password-file=.password --new-vault-password-file=.newpassword rekey secret.yaml
```

We can now use the `cat` command on the `secret.yaml` file and we will see the following output:

```
$ANSIBLE_VAULT;1.1;AES256
326234663566396466613261643139653133663939356232363234653562653136303539303
46135
373061643335333137653734396236666161636338623533306134326130313233643761346 46
36332
366565646538366162383838383638356263330373765333761356630343162633237646565316
56137
3462323739653339360a6139336338653838373933331616363653765646165363333303232 6
33132
633932373832313937383164653566363961333306132303932396263333735643230316361 3
83339
336539343836365306463663361663538653761393933361396539
```

This is very different than the previous one we had.

Vaults and playbooks

You can also use vaults with `ansible-playbook`. You'll need to decrypt the file on-the-fly using a command such as the following:

```
$ ansible-playbook site.yml --vault-password-file .password
```

There is yet another option that allows you to decrypt files using a script, which can then look up some other source and decrypt the file. This can also be a useful option to provide more security. However, make sure that the `get_password.py` script has executable permissions:

```
$ ansible-playbook site.yml --vault-password-file ~/.get_password.py
```

Before closing this chapter, I'd like to speak a little bit about the password file. This file needs to be present on the machine where you execute your playbooks, in a location and with permissions so that it is readable by the user who is executing the playbook. You can create the `.password` file at startup.

The `.` character in the `.password` filename is to make sure that the file is hidden by default when you look for it. This is not directly a security measure, but it could help mitigate cases where an attacker does not know exactly what they are looking for.

The `.password` file content should be a password or key that is secure and accessible only to folks who have permission to run Ansible playbooks.

Finally, make sure that you're not encrypting every file that's available! Ansible Vault should be used only for important information that needs to be secure.

Encrypting user passwords

Ansible Vault takes care of passwords that are checked in and helps you to handle them while running Ansible playbooks or commands. However, when Ansible plays are run, at times, you might need your users to enter passwords. You also want to make sure that these passwords don't appear in the comprehensive Ansible logs (the default `/var/log/ansible.log` location) or on `stdout`.

Ansible uses `Passlib`, which is a password-hashing library for Python, to handle encryption for prompted passwords. You can use any of the following algorithms supported by `Passlib`:

- `des_crypt`: DES crypt
- `bsdi_crypt`: BSDi crypt
- `bigcrypt`: BigCrypt
- `crypt16`: Crypt16
- `md5_crypt`: MD5 crypt
- `bcrypt`: BCrypt
- `sha1_crypt`: SHA-1 crypt
- `sun_md5_crypt`: Sun MD5 crypt
- `sha256_crypt`: SHA-256 crypt
- `sha512_crypt`: SHA-512 crypt
- `apr_md5_crypt`: Apache's MD5-crypt variant
- `phpass`: PHPass portable hash
- `pbkdf2_digest`: Generic PBKDF2 hashes

- `cta_pbkdf2_sha1`: Cryptacular's PBKDF2 hash
- `dlitz_pbkdf2_sha1`: Dwayne Litzenberger's PBKDF2 hash
- `scram`: SCRAM hash
- `bsd_nthash`: FreeBSD's MCF-compatible `nthash` encoding

Let's now see how encryption works with a variable prompt:

```
- name: ssh_password
  prompt: Enter ssh_password
  private: True
  encryption: md5_crypt
  confirm: True
  salt_size: 7
```

In the preceding snippet, `vars_prompt` is used to prompt users for some data.

This will prompt the user for the password, similarly to how SSH does it.

The `name` key indicates the actual variable name where Ansible will store the user password, as shown in the following command:

```
name: ssh_password
```

We are using the `prompt` key to prompt users for the password as follows:

```
prompt: Enter ssh password
```

We are explicitly asking Ansible to hide the password from `stdout` by using the `private` key; this works like any other password prompt on a Unix system. The `private` key is accessed as follows:

```
private: True
```

We are using the `md5_crypt` algorithm over here with a salt size of 7:

```
encrypt: md5_crypt
salt_size: 7
```

Moreover, Ansible will prompt for the password twice and compare both passwords:

```
confirm: True
```

Hiding passwords

Ansible, by default, filters output that contains the `login_password` key, the `password` key, and the `user:pass` format. For example, if you are passing a password in your module using `login_password` or the `password` key, then Ansible will replace your password with `VALUE_HIDDEN`. Let's now see how you can hide a password using the `password` key:

```
- name: Running a script
  shell: script.sh
    password: my_password
```

In the preceding `shell` task, we use the `password` key to pass passwords. This will allow Ansible to hide it from `stdout` and its log file.

Now, when you run the preceding task in the *verbose* mode, you should not see your `mypass` password; instead Ansible, with `VALUE_HIDDEN`, will replace it as follows:

```
REMOTE_MODULE command script.sh password=VALUE_HIDDEN #USE_SHELL
```

Using no_log

Ansible will hide your passwords only if you are using a specific set of keys. However, this might not be the case every time; moreover, you might also want to hide some other confidential data. The `no_log` feature of Ansible will hide your entire task from logging it to the `syslog` file. It will still print your task on `stdout` and log it to other Ansible log files.

 At the time of writing this book, Ansible did not support hiding tasks from `stdout` using `no_log`.

Let's now see how you can hide an entire task with `no_log`:

```
- name: Running a script
  shell: script.sh
    password: my_password
  no_log: True
```

By passing `no_log: True` to your task, Ansible will prevent the entire task from hitting `syslog`.

Summary

In this chapter, we have seen a very large number of Ansible features. We started with `local_actions` for performing operations on a machine, and then we moved on to delegation for performing the task on a third machine. We then moved on to conditionals and include for making playbooks more flexible. We learned about roles and how they can help you to keep your system aligned, and we learned how to organize an Ansible repository properly, making the most of Ansible and Git. Then, we covered execution strategies and Jinja filters for more flexible executions.

We ended this chapter with Ansible Vault and many other tips to make your Ansible execution safer.

In the next chapter, we will be looking at how to use Ansible to create infrastructures and, more specifically, how to do it using the cloud providers, AWS and DigitalOcean.

5

Going Cloud

In this chapter, we will see how to use Ansible for provisioning infrastructures in a matter of minutes. In my opinion, this is one of the most interesting and powerful capabilities of Ansible, since it allows you to recreate environments in a quick and consistent way. This is very important when you have multiple environments for the various stages of your deployment pipeline. In fact, it allows you to create equal environments and to keep them aligned when you need to make changes without any pain.

Letting Ansible provision your machines also has other advantages, and for those reasons I always suggest doing the following:

- **Audit trail**: In the last few years, the IT sector has swallowed a huge number of other sectors, and, as a consequence of this, the auditing processes are now looking at IT as a critical part of the process. When an auditor comes to the IT department asking for the history of a server, from its creation to the present moment, having Ansible playbooks for the whole process helps a lot.
- **Multiple staging environments**: As we mentioned before, if you have multiple environments, provisioning servers with Ansible will help you a lot.
- **Moving servers**: When a company uses a global cloud provider (such as AWS, DigitalOcean, or Azure), they often choose the region closest to their offices or customers at the moment they create the first servers. Those providers often open new regions, and if their new region is closer to you, you may want to move your whole infrastructure to the new region. This would be a nightmare if you had provisioned every resource manually!

In this chapter, on a broad level, we'll cover the following topics:

- Provisioning machines in AWS
- Provisioning machines in DigitalOcean
- Provisioning machines in Azure

Most of the new machine creations have two phases:

- Provisioning a new machine or a new set of machines
- Running playbooks to ensure the new machines are configured properly to play their role in your infrastructure

We've looked at the configuration management aspect in the initial chapters. We'll focus a lot more on provisioning new machines in this chapter, with a lesser focus on configuration management.

Technical requirements

You can download all the files from this book's GitHub repository at `https://github.com/PacktPublishing/Learning-Ansible-2.X-Third-Edition/tree/master/Chapter05`.

Provisioning resources in the cloud

With that, let's jump to the first topic. Teams managing infrastructures have a lot of choices today for running their builds, tests, and deployments. Providers such as Amazon, Azure, and DigitalOcean primarily provide **Infrastructure as a Service (IaaS)**. When we speak about IaaS, it's better to speak about resources not virtual machines for different reasons:

- The majority of the products that those companies allow you to provision are not machines but other critical resources such as networking and storage.
- Lately, many of those companies have started to provide many different kind of compute instances ranging from bare-metal machines to containers.
- Setting up machines with no networking (or storage) could be all you need for some very simple environments, but this might not be enough in production environments.

Those companies usually provide API, CLI, GUI, and SDK utilities to create and manage cloud resources throughout their whole life cycle. We're more interested in using their SDK as it will play an important part in our automation effort. Setting up new servers and provisioning them is interesting at first but at some stage it can become boring as it's quite repetitive in nature. Each provisioning step will involve several similar steps to get them up and running.

Imagine one fine morning you receive an email asking for three new customer setups, where each customer setup has three to four instances and a bunch of services and dependencies. This might be an easy task for you, but it would require running the same set of repetitive commands multiple times, followed by monitoring the servers once they come up to confirm that everything went well. In addition, anything you do manually has a chance of introducing problems. What if two of the customer setups come up correctly but, due to fatigue, you miss out a step for the third customer and hence introduce a problem? To deal with such situations, there exists automation.

Cloud-provisioning automation makes it easy for an engineer to build up a new server as quickly as possible, allowing them to concentrate on other priorities. Using Ansible, you can easily perform these actions and automate cloud provisioning with minimal effort. Ansible provides you with the power to automate various different cloud platforms, such as Amazon, Azure, DigitalOcean, Google Cloud, Rackspace, and many more, with modules for different services available in the Ansible core or extended module packages.

 As mentioned earlier, bringing up new machines is not the end of the game. We also need to make sure we configure them to play the required role.

In the next sections, we will provision the environment that we have used in the previous chapters (two web servers and one database server) in the following environments:

- **Simple AWS deployment**: Where all machines will be placed in the same **Availability Zones** (**AZs**) and the same network.
- **Complex AWS deployment**: Where the machines will be split in multiple AZs as well as networks.
- **DigitalOcean**: DigitalOcean does not allow us to do many networking tweaks so it will be similar to the first one.
- **Azure**: We will create a simple deployment in this case.

Provisioning machines in AWS

AWS is the most used public cloud by a fair amount, and it's often chosen due to its huge number of available services as well as the huge amount of documentation, answered questions, and articles that can be expected around such a popular product.

Since AWS's goal is to be a complete virtual data center provider (and much more), we will need to create and manage our network as we would do if we had to set up a real data center. Obviously, we will not need to cable stuff, since it's a virtual data center. Due to this, a few lines of an Ansible playbook will be enough.

AWS global infrastructure

Amazon has always been pretty discreet about sharing the location or the exact number of data centers that its cloud is actually composed of. While I'm writing this, AWS counts 21 regions (with four more regions already announced) with a total of 61 AZs and hundreds of edge locations. Amazon defines a region as a *"physical location in the world where we (Amazon) have multiple AZs"*. Looking at Amazon's documentation of AZs, it says that *"an AZ consists of one or more discrete data centers, each with redundant power, networking, and connectivity, housed in separate facilities"*. For edge location, there is no official definition.

As you can see, from a real-life point of view, those definitions do not help you much. When I try to explain those concepts, I usually use different definitions, created by myself:

- **Region**: A group of AZs that are physically close
- **AZ**: A data center in a region (Amazon says that it could be more than one data center, but since there is no document listing the specific geometry of every AZ, I assume the worst-case scenario)
- **Edge location**: Internet exchanges or third-party data centers where Amazon has CloudFront and Route 53 endpoints

Even though I tried to make those definitions as easy and as useful as possible, some of them are very cloudy. When we start to speak about real-world differences, the definitions will become immediately clear. For instance, from a network speed perspective, when you move content within the same AZ, the bandwidth is very high. When you do the same operation with two AZs in the same region, you get high bandwidth, while if you use two AZs from two different regions, the bandwidth will be much lower. Also, there is a price difference, since all traffic within the same region is free, while traffic among different regions is free of charge.

AWS Simple Storage Service

Amazon **Simple Storage Service** (**S3**) is the first AWS service to be launched and it's also one of the most well-known AWS services. Amazon S3 is an object storage service with public endpoints as well as private endpoints. It uses the concept of a bucket to allow you different kinds of files and to manage them in a simple way. Amazon S3 also gives the user more advanced features such as the capability of serving a bucket's contents using a built-in web server. This is one of the reasons why many people decide to host their website, or the pictures on their websites, on Amazon S3.

The advantages of S3 are mainly the following:

- **Price schema**: You are billed by used gigabyte/month and by gigabyte transferred.
- **Reliability**: Amazon affirms that the objects on AWS S3 have a 99.999999999% probability to survive any given year. This is orders of magnitude higher than any hard disk.
- **Tooling**: Since S3 is a service that has been out there for many years now, a lot of tools have been implemented to leverage this service.

AWS Elastic Compute Cloud

The second service launched by AWS is the **Elastic Compute Cloud** (**EC2**) service. This service allows you to spin up machines on AWS infrastructure. You can think of those EC2 instances as OpenStack compute instances or VMware virtual machines. Initially, those machines were very similar to VPS, but after a while Amazon decided to give much more flexibility on those machines, introducing a very advanced networking option. The old kind of machines are still available in the oldest data centers with the name EC2 Classic, while the new kind is the current default and is just called EC2.

AWS Virtual Private Cloud

Virtual Private Cloud (**VPC**) is Amazon's networking implementation that we mentioned in the previous section. The VPC is more a set of tools than a single tool; in fact, the capabilities it offers were offered by multiple metal boxes in the classic data center.

The main things you can create with VPC are the following:

- Switches
- Routers
- DHCP
- Gateways
- Firewalls
- **Virtual Private Networks (VPNs)**

An important thing to understand when you use VPC is that the layout of your network is not completely arbitrary, since Amazon has created a few limitations to simplify their networking. The basic limitations are these:

- You cannot spawn a subnetwork between AZs.
- You cannot spawn a network between regions.
- You cannot route networks in different regions directly.

While, for the first two, the only solution is creating multiple networks and subnetworks, for the third, you can actually implement a workaround using a VPN service that could be self-provisioned or be provisioned using the official AWS VPN service.

We will be mainly using the switching and routing capabilities of VPC.

AWS Route 53

Like many other cloud services, Amazon offers a **DNS as a Service (DNSaaS)** feature and in Amazon case it's called **Route 53**. Route 53 is a distributed DNS service with hundreds of endpoints worldwide (Route 53 is present in all AWS edge locations).

Route 53 allows you to create different zones for a domain allowing split-horizon situations in which based on the fact that the client asking for a DNS resolution is inside or outside your VPC will receive different responses. This is very useful when you want your applications to be easily moved in and out of your VPC without changes, but at the same time you want your traffic to stay on a private (virtual) network whenever possible.

AWS Elastic Block Storage

AWS **Elastic Block Storage** (**EBS**) is a block storage provider for allowing your EC2 instances to keep data that will survive reboots and is very flexible. From a user perspective, EBS seems a lot like any other SAN product with a simpler interface, since you only need to create the volume and tell EBS to which machine it needs to be attached, and EBS does the rest. You can attach multiple volumes to a single server, but every volume can be connected to only one server at any given time.

AWS Identity and Access Management

To allow you to manage users and access methods, Amazon provides the **Identity and Access Management** (**IAM**) service. The main features of the IAM service are as follows:

- Create, edit, and delete users
- Change user password
- Create, edit, and delete groups
- Manage users and group association
- Manage tokens
- Manage two-factor authentication
- Manage SSH keys

We will be using this service to set up our users and their permissions.

Amazon Relational Database Service

Setting up and maintaining relational databases is complex and very time-consuming. To simplify this, Amazon provides some widely used **Database as a Service** (**DBaaS**), more specifically, the following:

- Aurora
- MariaDB
- MySQL
- Oracle
- PostgreSQL
- SQL Server

For each one of those engines, Amazon offers different features and price models but the specifics of each is beyond the goal of this book.

Setting up an account with AWS

The first thing we will need before starting to work on our AWS is an account. Creating an account on AWS is pretty straightforward and very well documented by Amazon official documentation as well as by multiple independent sites, and therefore it will not be covered in these pages.

After you have created your AWS account, you need to go in to the AWS and do the following:

- Upload your SSH key in **EC2** | **Keypairs.**
- Create a new user in **Identity & Access Management** | **Users** | **Create new user** and create a file in ~/.aws/credentials with the following lines:

```
[default]
aws_access_key_id = YOUR_ACCESS_KEY
aws_secret_access_key = YOUR_SECRET_KEY
```

After you have created your AWS keys and uploaded your SSH key, you need to set up Route 53. In Route 53, you need to create two zones for your domain (you can also use a subdomain, if you don't have an unused domain): one public and one private.

If you create only the public zone, Route 53 will propagate this zone everywhere, but if you create a public and a private zone, Route 53 will serve your public zone everywhere but in the VPC you specified when creating the private zone. If you query those DNS entries from within that VPC, the private zone will be used. This approach has multiple advantages:

- Only publicize the IP addresses of public machines
- Always use DNS names instead of IP addresses, even for internal traffic
- Ensure that your internal machines communicate directly without your traffic ever passing through the public web
- Since the external IPs in AWS are virtual IPs managed by Amazon and associated to your instances using NATs, this approach grants the least amount of hops and therefore latency

If you declared an entry for your public zone but not in the private one, the machines in the VPC will not be able to resolve that entry.

After you have created the public zone, AWS will give you a few name server IP addresses and you need to put those in your register/root zone DNS so that you can actually resolve those DNS.

Simple AWS deployment

As we said previously, the first thing that we will need is the networking up. For this example, we will need just one single network in one AZ and all our machines will stay there.

In this section, we will be working in the `playbooks/aws_simple_provision.yaml` file.

The first two lines are just used to declare the host that will perform the commands (localhost) and the beginning of the tasks section:

```
- hosts: localhost
  tasks:
```

First, we are going to ensure that the public/private key pair is present:

```
- name: Ensure key pair is present
  ec2_key:
    name: fale
    key_material: "{{ lookup('file', '~/.ssh/fale.pub') }}"
```

In AWS, we need to have a VPC network and subnetwork. By default, they are already there, but if you need it, you can do the following to create the VPC network:

```
- name: Ensure VPC network is present
  ec2_vpc_net:
    name: NET_NAME
    state: present
    cidr_block: 10.0.0.0/16
    region: AWS_REGION
  register: aws_net
- name: Ensure the VPC subnetwork is present
  ec2_vpc_subnet:
    state: present
    az: AWS_AZ
    vpc_id: '{{ aws_simple_net.vpc_id }}'
    cidr: 10.0.1.0/24
  register: aws_subnet
```

Since we are using the default VPC, we need to query AWS to know the VPC network and subnetwork values:

```
  - name: Ensure key pair is present
    ec2_key:
      name: fale
      key_material: "{{ lookup('file', '~/.ssh/fale.pub') }}"
  - name: Gather information of the EC2 VPC net in eu-west-1
ec2_vpc_net_facts:
region: eu-west-1
register: aws_simple_net
  - name: Gather information of the EC2 VPC subnet in eu-west-1
ec2_vpc_subnet_facts:
region: eu-west-1
filters:
vpc-id: '{{ aws_simple_net.vpcs.0.id }}'
register: aws_simple_subnet
```

Now we have all the information we need on the network and subnetwork, we can move to security groups. We can do this with the `ec2_group` module. In the AWS world, security groups are used for firewalling. Security groups are very similar to groups of firewall rules that share the same destination (for ingress rules) or the same destination (for egress rules). Three differences with standard firewalls rules are actually worth mentioning:

- Multiple security groups can be applied to the same EC2 instance.
- As a source (for ingress rules) or destination (for egress rules), you can specify one of the following:
  - An instance ID
  - Another security group
  - An IP range
- You don't have to specify a default deny rule at the end of the chain, because AWS will add it by default.

So, in my case, the following code will be added to `playbooks/aws_simple_provision.yaml`:

```
  - name: Ensure wssg Security Group is present
    ec2_group:
      name: wssg
      description: Web Security Group
      region: eu-west-1
      vpc_id: '{{ aws_simple_net.vpcs.0.id }}'
      rules:
        - proto: tcp
          from_port: 22
```

```
            to_port: 22
            cidr_ip: 0.0.0.0/0
          - proto: tcp
            from_port: 80
            to_port: 80
            cidr_ip: 0.0.0.0/0
          - proto: tcp
            from_port: 443
            to_port: 443
            cidr_ip: 0.0.0.0/0
        rules_egress:
          - proto: all
            cidr_ip: 0.0.0.0/0
      register: aws_simple_wssg
```

We are now going to create another security group for our database. In this case, we only need to open port 3036 to the servers in the web security group:

```
    - name: Ensure dbsg Security Group is present
      ec2_group:
        name: dbsg
        description: DB Security Group
        region: eu-west-1
        vpc_id: '{{ aws_simple_net.vpcs.0.id }}'
        rules:
          - proto: tcp
            from_port: 3036
            to_port: 3036
            group_id: '{{ aws_simple_wssg.group_id }}'
        rules_egress:
          - proto: all
            cidr_ip: 0.0.0.0/0
```

 As you can see, we allow all egress traffic to flow. This is not what security best practices suggest, and therefore you may need to regulate egress traffic as well. A case that frequently forces you to regulate egress traffic is if you want your target machine to be PCI-DSS compliant.

Now that we have the VPC, the subnet into the VPC, and the needed security groups, we can now move on to actually creating the EC2 instances:

```
- name: Setup instances
  ec2:
    assign_public_ip: '{{ item.assign_public_ip }}'
    image: ami-3548444c
    region: eu-west-1
    exact_count: 1
    key_name: fale
    count_tag:
      Name: '{{ item.name }}'
    instance_tags:
      Name: '{{ item.name }}'
    instance_type: t2.micro
    group_id: '{{ item.group_id }}'
    vpc_subnet_id: '{{ aws_simple_subnet.subnets.0.id }}'
    volumes:
      - device_name: /dev/sda1
        volume_type: gp2
        volume_size: 10
        delete_on_termination: True
  register: aws_simple_instances
  with_items:
    - name: ws01.simple.aws.fale.io
      group_id: '{{ aws_simple_wssg.group_id }}'
      assign_public_ip: True
    - name: ws02.simple.aws.fale.io
      group_id: '{{ aws_simple_wssg.group_id }}'
      assign_public_ip: True
    - name: db01.simple.aws.fale.io
      group_id: '{{ aws_simple_dbsg.group_id }}'
      assign_public_ip: False
```

 When we created the DB machine, we did not specify the `assign_public_ip: True` line. In this case, the machine will not receive a public IP, and therefore it will not be reachable from outside our VPC. Since we used a very strict security group for this server, it would not be reachable from any machine outside the `wssg` anyway.

As you can guess, the piece of code we have just seen will create our three instances (two web servers and one DB server).

We can now proceed to add those newly created instances to our Route 53 account so that we can resolve those machines' FQDN. To interact with AWS Route 53, we will be using the `route53` module, which allows us to create entries, query entries, and delete entries. To create a new entry, we will be using the following code:

```
- name: Add route53 entry for server SERVER_NAME
  route53:
    command: create
    zone: ZONE_NAME
    record: RECORD_TO_ADD
    type: RECORD_TYPE
    ttl: TIME_TO_LIVE
    value: IP_VALUES
    wait: True
```

So, to create the entries for our servers, we will add the following code:

```
- name: Add route53 rules for instances
  route53:
    command: create
    zone: aws.fale.io
    record: '{{ item.tagged_instances.0.tags.Name }}'
    type: A
    ttl: 1
    value: '{{ item.tagged_instances.0.public_ip }}'
    wait: True
  with_items: '{{ aws_simple_instances.results }}'
  when: item.tagged_instances.0.public_ip
- name: Add internal route53 rules for instances
  route53:
    command: create
    zone: aws.fale.io
    private_zone: True
    record: '{{ item.tagged_instances.0.tags.Name }}'
    type: A
    ttl: 1
    value: '{{ item.tagged_instances.0.private_ip }}'
    wait: True
  with_items: '{{ aws_simple_instances.results }}'
```

 Since the database server does not have a public address, it makes no sense to publish this machine in the public zone, so we have created this machine entry only in the internal zone.

Putting it all together, the `playbooks/aws_simple_provision.yaml` will be the following. The full code is available on GitHub:

```
---
- hosts: localhost
  tasks:
    - name: Ensure key pair is present
      ec2_key:
        name: fale
        key_material: "{{ lookup('file', '~/.ssh/fale.pub') }}"
    - name: Gather information of the EC2 VPC net in eu-west-1
      ec2_vpc_net_facts:
        region: eu-west-1
      register: aws_simple_net
    - name: Gather information of the EC2 VPC subnet in eu-west-1
      ec2_vpc_subnet_facts:
        region: eu-west-1
        filters:
          vpc-id: '{{ aws_simple_net.vpcs.0.id }}'
      register: aws_simple_subnet
    ...
```

Running it with `ansible-playbook playbooks/aws_simple_provision.yaml`, Ansible will take care of creating our environment.

Complex AWS deployment

In this section, we will slightly change the previous example to move one of the web servers to another AZ within the same region. To do so, we are going to make a new file in `playbooks/aws_complex_provision.yaml` that will be very similar to the previous one, with one difference located in the part that helps us provision the machines. In fact, we will use the following code snippet instead of the one we used on the previous run. The full code is available on GitHub:

```
    - name: Setup instances
      ec2:
        assign_public_ip: '{{ item.assign_public_ip }}'
        image: ami-3548444c
        region: eu-west-1
        exact_count: 1
        key_name: fale
        count_tag:
          Name: '{{ item.name }}'
        instance_tags:
          Name: '{{ item.name }}'
```

```
instance_type: t2.micro
group_id: '{{ item.group_id }}'
vpc_subnet_id: '{{ item.vpc_subnet_id }}'
volumes:
  - device_name: /dev/sda1
    volume_type: gp2
    volume_size: 10
    delete_on_termination: True
...
```

As you can see, we have put the `vpc_subnet_id` in a variable, so that we can use a different one for the `ws02` machine. Due to the fact that AWS already provides two subnets by default (and every subnet is tied to a different AZ), it's enough to use the following AZ. Security groups and Route 53 code does not need to be changed, since it does not work at a subnet/AZ level, but at a VPC level (for security groups and the internal Route 53 zone) or a global level (for public Route 53).

Provisioning machines in DigitalOcean

Compared to AWS, DigitalOcean seems to be very incomplete. DigitalOcean, until a few months ago, only provided droplets, SSH key management, and DNS management. At the time of writing, DigitalOcean has very recently launched an additional block storage service. The advantages of DigitalOcean compared to many competitors are as follows:

- Lower prices than AWS.
- Very easy APIs.
- Very well-documented APIs.
- The droplets are very similar to standard virtual machines (they don't do weird customization).
- The droplets are very quick to go up and down.
- Since DigitalOcean has a very simple networking stack, it's way more efficient than the AWS one.

Droplets

Droplets are the main service offered by DigitalOcean and are compute instances that are very similar to Amazon EC2 Classic. DigitalOcean relies on the **Kernel Virtual Machine** (**KVM**) to virtualize the machines, assuring very high performance and security.

Since they do not change KVM in any sensible way, and since KVM is open source and available on any Linux machine, this allows system administrators to create identical environments on private and public clouds. DigitalOcean droplets will have one external IP and they can be eventually added to a virtual network that will allow your machines to use internal IPs.

Different than many other comparable services, DigitalOcean allows your droplets to have IPv6 IPs in addition to the IPv4 ones. This service is free of charge.

SSH key management

Every time you want to create a droplet, you have to specify whether you want a specific SSH key assigned to the root user or if you want a password (which will have to be changed at the first login). To be able to choose an SSH key, you need an interface to upload it. DigitalOcean allows you to do this using a very simple interface which allows you to list the current keys, as well as create and delete keys.

Private networking

As mentioned in the droplet section, DigitalOcean allows us to have a private network where our machine can communicate with another. This allows segregation of services (such as a database service) only on the internal network to allow a higher level of security. Since, by default, MySQL binds on all available interfaces, we will need to tweak the database role a little bit to only bind on the internal network.

To recognize the internal network from the external one, there are many ways, due to some DigitalOcean peculiarities:

- Private networks are always in the `10.0.0.0/8` network, while public IPs are never in that network.
- The public network is always `eth0`, while the private network is always `eth1`.

Based on your portability needs, you can use either one of those strategies to understand where to bind your services.

Adding an SSH key in DigitalOcean

We firstly need a DigitalOcean account. As soon as we have a DigitalOcean user, a credit card set up, and the API key, we can start to use Ansible to add our SSH key to our DigitalOcean cloud. To do so, we need to create a file called `playbooks/do_provision.yaml` with the following structure:

```
- hosts: localhost
  tasks:
    - name: Add the SSH Key to Digital Ocean
      digital_ocean:
        state: present
        command: ssh
        name: SSH_KEY_NAME
        ssh_pub_key: 'ssh-rsa AAAA...'
        api_token: XXX
      register: ssh_key
```

In my case, this is my file content:

```
    - name: Add the SSH Key to Digital Ocean
      digital_ocean:
        state: present
        command: ssh
        name: faleKey
        ssh_pub_key: "{{ lookup('file', '~/.ssh/fale.pub') }}"
        api_token: ee02b...2f11d
      register: ssh_key
```

Then we can execute it and you will have a result similar to the following:

```
PLAY [all] *************************************************************

TASK [Gathering Facts] ************************************************
ok: [localhost]

TASK [Add the SSH Key to Digital Ocean] *****************************
changed: [localhost]

PLAY RECAP ************************************************************
localhost : ok=2 changed=1 unreachable=0 failed=0
```

This task is idempotent so we can execute it multiple times. If the key has already been uploaded, the SSH key ID will be returned at every run.

Deployment in DigitalOcean

 At the time of writing, the only way to create a droplet in Ansible is by using the `digital_ocean` module that could be soon deprecated since many of its features are now done in a better, cleaner way by other modules and there is already a bug on the Ansible bug tracker to track its complete rewrite and possible deprecation. My guess is that the new module will be called `digital_ocean_droplet` and will have a similar syntax, but at the moment there is no code so it's just my guess.

To create the droplets, we will have to use the `digital_ocean` module with a syntax similar to the following:

```
- name: Ensure the ws and db servers are present
  digital_ocean:
    state: present
    ssh_key_ids: KEY_ID
    name: '{{ item }}'
    api_token: DIGITAL_OCEAN_KEY
    size_id: 512mb
    region_id: lon1
    image_id: centos-7-0-x64
    unique_name: True
  with_items:
    - WEBSERVER 1
    - WEBSERVER 2
    - DBSERVER 1
```

To make sure that all our provisioning is done completely and in a sane way, I always suggest creating one single provision file for the whole infrastructure. So, in my case, I'll add the following task to the `playbooks/do_provision.yaml` file:

```
- name: Ensure the ws and db servers are present
  digital_ocean:
    state: present
    ssh_key_ids: '{{ ssh_key.ssh_key.id }}'
    name: '{{ item }}'
    api_token: ee02b...2f11d
    size_id: 512mb
    region_id: lon1
    image_id: centos-7-x64
```

```
          unique_name: True
        with_items:
          - ws01.do.fale.io
          - ws02.do.fale.io
          - db01.do.fale.io
        register: droplets
```

After this, we can add the domain with the `digital_ocean_domain` module:

```
      - name: Ensure domain resolve properly
        digital_ocean_domain:
          api_token: ee02b...2f11d
          state: present
          name: '{{ item.droplet.name }}'
          ip: '{{ item.droplet.ip_address }}'
        with_items: '{{ droplets.results }}'
```

So, putting all this together, our `playbooks/do_provision.yaml` will look like this, and the full code block is available on GitHub:

```
---
- hosts: localhost
  tasks:
    - name: Ensure domain is present
      digital_ocean_domain:
        api_token: ee02b...2f11d
        state: present
        name: do.fale.io
        ip: 127.0.0.1
    - name: Add the SSH Key to Digital Ocean
      digital_ocean:
        state: present
        command: ssh
        name: faleKey
        ssh_pub_key: "{{ lookup('file', '~/.ssh/fale.pub') }}"
        api_token: ee02b...2f11d
      register: ssh_key
  ...
```

So, we can now run it with the following command:

```
ansible-playbook playbooks/do_provision.yaml
```

We will see a result similar to the following. Full code output file is available on GitHub:

```
PLAY [localhost] *********************************************

TASK [Gathering Facts] *********************************************
```

```
ok: [localhost]

TASK [Ensure domain is present] *************************************
changed: [localhost]

TASK [Add the SSH Key to Digital Ocean] ****************************
changed: [localhost]

TASK [Ensure the ws and db servers are present] *******************
changed: [localhost] => (item=ws01.do.fale.io)
changed: [localhost] => (item=ws02.do.fale.io)
changed: [localhost] => (item=db01.do.fale.io)


...
```

We have seen how to provision three machines on DigitalOcean with few lines of Ansible. We can now configure them with the playbooks we have discussed in the previous chapters.

Provisioning machines in Azure

Lately, Azure is becoming one of the biggest clouds, mainly in some companies.

As you may imagine, Ansible has Azure-specific modules to create Azure environments without pain.

The first thing we will need on Azure, after having created the account, is to set up the authorization.

There are several ways to do it, but the easiest one is probably creating the `~/.azure/credentials` file in the INI format containing a `[default]` section with `subscription_id` and, alternatively, `client_id` and `secret` or `ad_user` and `password`.

An example of this would be the following file:

```
[default]
subscription_id: __AZURE_SUBSCRIPTION_ID__
client_id: __AZURE_CLIENT_ID__
secret: __AZURE_SECRET__
```

After this, we need a resource group in which we then will create all our resources.

To do so, we can use the `azure_rm_resourcegroup`, with the following syntax:

```
- name: Create resource group
  azure_rm_resourcegroup:
    name: myResourceGroup
    location: eastus
```

Now that we have the resource group, we can create a virtual network and a virtual subnetwork into it:

```
- name: Create Azure VM
      hosts: localhost
      tasks:
  - name: Create resource group
      azure_rm_resourcegroup:
      name: myResourceGroup
      location: eastus
  - name: Create virtual network
      azure_rm_virtualnetwork:
      resource_group: myResourceGroup
      name: myVnet
      address_prefixes: "10.0.0.0/16"
- name: Add subnet
      azure_rm_subnet:
      resource_group: myResourceGroup
    name: mySubnet
    address_prefix: "10.0.1.0/24"
    virtual_network: myVnet
```

Before we can progress with the creation of the virtual machine, we still need some networking items, and more specifically, a public IP, a network security group, and a virtual network card:

```
- name: Create public IP address
  azure_rm_publicipaddress:
    resource_group: myResourceGroup
    allocation_method: Static
    name: myPublicIP
  register: output_ip_address
- name: Dump public IP for VM which will be created
  debug:
    msg: "The public IP is {{ output_ip_address.state.ip_address }}."
- name: Create Network Security Group that allows SSH
  azure_rm_securitygroup:
    resource_group: myResourceGroup
    name: myNetworkSecurityGroup
    rules:
      - name: SSH
```

```
            protocol: Tcp
            destination_port_range: 22
            access: Allow
            priority: 1001
            direction: Inbound
    - name: Create virtual network inteface card
      azure_rm_networkinterface:
        resource_group: myResourceGroup
        name: myNIC
        virtual_network: myVnet
        subnet: mySubnet
        public_ip_name: myPublicIP
        security_group: myNetworkSecurityGroup
```

Now we are ready to create our first Azure machine, with the following code:

```
    - name: Create VM
      azure_rm_virtualmachine:
        resource_group: myResourceGroup
        name: myVM
        vm_size: Standard_DS1_v2
        admin_username: azureuser
        ssh_password_enabled: false
        ssh_public_keys:
          - path: /home/azureuser/.ssh/authorized_keys
            key_data: "{{ lookup('file', '~/.ssh/fale.pub') }}"
        network_interfaces: myNIC
        image:
          offer: CentOS
          publisher: OpenLogic
          sku: '7.5'
        version: latest
```

Running the playbook, you'll obtain a CentOS machine running on Azure!

Summary

In this chapter, we have seen how we can provision our machines in the AWS cloud, the DigitalOcean one, and Azure. In the case of the AWS cloud, we have seen two different examples, one very simple and one slightly more complex.

In the next chapter, we will talk about getting notified by Ansible when something goes south.

6
Getting Notification from Ansible

One of the big advantages of Ansible, compared to a bash script, is its idempotency, ensuring that everything is in order. This is a very nice feature that not only assures you that nothing has changed the configurations on your server, but also that new configurations will be applied in a short time.

Due to these reasons, many people run their `master.yaml` file once a day. When you do this (and probably you should!), you want some kind of feedback sent to you by Ansible itself. There are also many other cases where you may want Ansible to send messages to you or your team. For instance, if you use Ansible to deploy your application, you may want to send an IRC message (or other kinds of group chat messages) to your development team channel, so that they are all informed of the status of your system.

Other times, you want Ansible to notify Nagios that it's going to break something, so that Nagios does not worry and does not start to shoot emails and messages to your system administrators. In this chapter, we are going to look at multiple ways to help you to set up Ansible Playbooks that can both work with your monitoring system and eventually send notifications.

In this chapter, we'll explore the following topics:

- Email notifications
- Ansible XMPP/Jabber
- Slack and Rocket chat
- Sending a message to an IRC channel (community information and contributing)
- Amazon Simple Notification Service
- Nagios

Technical requirements

Many of the examples will require third-party systems (to send messages to), which you may or may not be using. If you do not have access to one of these systems, the relative example will not be executable for you. This is not a big problem, since you can still read the section and many notification modules are very similar to another. You might find that a module that is more suitable for your environment is present and functions in a very similar way. For a full list of Ansible notification modules, you can refer to `https://docs.ansible.com/ansible/latest/modules/list_of_notification_modules.html`.

You can download all of the files from this book's GitHub repository at: `https://github.com/PacktPublishing/Learning-Ansible-2.X-Third-Edition/tree/master/Chapter06`.

Sending emails with Ansible

It's frequent that a user needs to be promptly notified about an action that an Ansible Playbook performs. This can be either because it's important that this user knows about it, or because there is an automated system that has to be notified in order to (not) start some procedure.

The easiest and most common way of alerting people is to send emails. Ansible allows you to send emails from your playbook using a `mail` module. You can use this module in between any of your tasks and notify your user whenever required. Also, in some cases, you cannot automate each and every thing because either you lack the authority or it requires some manual checking and confirmation. If this is the case, you can notify the responsible user that Ansible has done its job and it's time for them to perform their duty. Let's use the `mail` module to notify your users with a very simple playbook called `uptime_and_email.yaml`:

```
---
- hosts: localhost
  connection: local
  tasks:
    - name: Read the machine uptime
      command: uptime -p
      register: uptime
    - name: Send the uptime via e-mail
      mail:
        host: mail.fale.io
        username: ansible@fale.io
        password: PASSWORD
        to: me@fale.io
```

```
subject: Ansible-report
body: 'Local system uptime is {{ uptime.stdout }}.'
```

The preceding playbook firstly reads the current machine uptime, issuing the `uptime` command, and then sends it via email to the `me@fale.io` email address. To send the email, we clearly need some additional information, such as the SMTP host, a valid set of SMTP credentials, and the content of the email. This example is very easy and will allow us to keep the examples short, but obviously, you can generate the emails in a similar way in very long and complex playbooks. If we focus on the `mail` task a little bit, we can see that we are using it with the following data:

- An email server to be used to send the email (also with login information, which is required for this server)
- The receiver email address
- The email subject
- The email body

Other interesting parameters that the `mail` module supports are the following:

- The `attach` parameter: This is used to add attachments to the email that will be generated. This is very useful when, for instance, you want to send a log via an email.
- The `port` parameter: This is used to specify which port is used by the email server.

An interesting thing about this module is that the only mandatory field is `subject`, and not the body, as many people would expect. The RFC 2822 does not force the presence of either the subject or the body, so an email without both of them is still valid, but it would be very hard for a human to manage such formatted email. Ansible, therefore, will always send emails with both subject and body, and if the body is empty, it will use the `subject` string both in the subject and in the body.

We can now proceed to execute the script to validate its functionality with the following command:

```
ansible-playbook -i localhost, uptime_and_email.yaml
```

 This playbook might not work on some machines, since the `-p` parameter of `uptime` is Linux-specific and might not work on other POSIX operating systems, such as macOS.

By running the previous playbook, we will have a result similar to the following:

```
PLAY [localhost] ************************************************
TASK [setup] ***************************************************
ok: [localhost]
TASK [Read the machine uptime] *******************************
changed: [localhost]
TASK [Send the uptime via email] *****************************
changed: [localhost]
PLAY RECAP ***************************************************
localhost           : ok=3    changed=2    unreachable=0    failed=0
```

Also, as expected, Ansible has sent me an email with the following content:

```
Local system uptime is up 38 min
```

This module can be used in many different ways. An example of a real-world case that I've seen is a playbook that was created to automate a piece of a very complex, but sequential process, where multiple people were involved. Each person had a specific point in the process in which they had to start their work and the next person in the chain cannot start their work before the previous person completed theirs. The key to keep the process ticking was each person emailing the next person in the chain to notify them that their own part was completed and therefore the receiver needed to start their work in the process. The people in the process generally carried out the email notification manually. When we started to automate that procedure, we did it for one specific piece and no one noticed that that part was automated.

Having such long processes tracked via email is not the best way to handle them since errors are easy to make, potentially losing track of the process. Also, those kind of complex and sequential processes tend to be very slow, but it's widely used in organizations and often you cannot change it.

There are cases where the process needs to send notifications in a more real-time way than emails, so XMPP could be a good way to do it.

XMPP

Emails are slow, unreliable, and often people do not react to them immediately. There are cases where you want to send a real-time message to one of your users. Many organizations rely on XMPP/Jabber for their internal chat system and the great thing is that Ansible is able to directly send messages to XMPP/Jabber users and conference rooms.

Let's tweak the previous example to send uptime information to a user in the
`uptime_and_xmpp_user.yaml` file:

```
---
- hosts: localhost
  connection: local
  tasks:
    - name: Read the machine uptime
      command: 'uptime -p'
      register: uptime
    - name: Send the uptime to user
      jabber:
        user: ansible@fale.io
        password: PASSWORD
        to: me@fale.io
        msg: 'Local system uptime is {{ uptime.stdout }}.'
```

 If you want to use the Ansible `jabber` task, you will need to have the
`xmpppy` library installed on the system that will perform the task. One
way to install it is by using your package manager. For instance, on
Fedora you can just execute `sudo dnf install -y python2-xmpp` and
it will be installed. You can also use `pip install xmpppy`.

The first task is exactly the same as we had in the previous section, while the second one
has small differences. As you can see, the `jabber` module is very similar to the `mail`
module and requires similar parameters. In the XMPP case, we don't need to specify the
server host and port, since that information is automatically gathered by XMPP from the
DNS. In cases where we would need to use a different server host or port, we can use,
respectively, the `host` and `port` parameters.

We can now proceed to execute the script to validate its functionality with the following
command:

```
ansible-playbook -i localhost, uptime_and_xmpp_user.yaml
```

We will have a result similar to the following:

```
PLAY [localhost] **********************************************
TASK [setup] **************************************************
ok: [localhost]
TASK [Read the machine uptime] *******************************
changed: [localhost]
TASK [Send the uptime to user] *******************************
changed: [localhost]
PLAY RECAP ***************************************************
localhost          : ok=3    changed=2    unreachable=0    failed=0
```

In cases where we want to send a message to a conference room instead of a single user, it is enough to change the receiver in the `to` parameter, by adding the associated conference room instead:

```
to: sysop@conference.fale.io (mailto:sysop@conference.fale.io)/ansiblebot
```

Except for the receiver change and the addition of `(mailto:sysop@conference.fale.io)/ansiblebot` which identifies the chat handle to use (`ansiblebot`, in this case), XMPP handles users and conference rooms in the same way, so it's very easy to switch from one to the other.

Although XMPP is fairly popular, not every company uses it. Another collaboration platform that Ansible can send messages to is Slack.

Slack

In the last few years, many new chat and collaboration platforms have appeared. One of the most used ones is Slack. Slack is a cloud-based team collaboration tool, and this allows even easier integration with Ansible than XMPP.

Let's put the following lines in the `uptime_and_slack.yaml` file:

```
---
- hosts: localhost
  connection: local
  tasks:
    - name: Read the machine uptime
      command: 'uptime -p'
      register: uptime
    - name: Send the uptime to slack channel
      slack:
        token: TOKEN
        channel: '#ansible'
        msg: 'Local system uptime is {{ uptime.stdout }}.'
```

As we discussed, this module has an even simpler syntax than the XMPP one. In fact, it only needs to know the token (which you can generate on the Slack website), the channel to send the message to, and the message itself.

 Since version 1.8 of Ansible, the new version of the Slack token is required, for instance, `G522SJP14/D563DW213/7Qws484asdWD4w12Md3avf4FeD`.

Run the playbook with the following:

```
ansible-playbook -i localhost, uptime_and_slack.yaml
```

This results in the following output:

```
PLAY [localhost] ***********************************************
TASK [setup] **************************************************
ok: [localhost]
TASK [Read the machine uptime] *********************************
changed: [localhost]
TASK [Send the uptime to slack channel] ***********************
changed: [localhost]
PLAY RECAP ****************************************************
localhost          : ok=3    changed=2    unreachable=0    failed=0
```

Since Slack's goal is to make communications more efficient, it allows us to tweak multiple aspects of the message. The most interesting points, from my point of view, are the following:

- `color`: This allows you to specify a color bar to be put at the beginning of the message to identify the following states:
  - Good: green bar
  - Normal: no bar
  - Warning: yellow bar
  - Danger: red bar
- `icon_url`: This allows you to change the user image for that message.

For example, the following code would send a message with a warning color and a custom user image:

```
- name: Send the uptime to slack channel
  slack:
    token: TOKEN
    channel: '#ansible'
    msg: 'Local system uptime is {{ uptime.stdout }}.'
    color: warning
    icon_url: https://example.com/avatar.png
```

Since not every company is comfortable with Slack being able to see their private conversations, there are alternatives, such as Rocket Chat.

Rocket Chat

Many companies like the functionality of Slack, but do not want to lose out on the privacy that an on-premises service gives you when using Slack. **Rocket Chat** is an open source software solution that implements most of the features of Slack, as well as the majority of its interface. Being open source, every company can install it on-premises and manage it in a way that is compliant with their IT rules.

As Rocket Chat's goal is to be a drop-in replacement for Slack, from our point of view, very few changes need to be done. In fact, we can create the `uptime_and_rocket.yaml` file with the following content:

```
---
- hosts: localhost
  connection: local
  tasks:
    - name: Read the machine uptime
      command: 'uptime -p'
      register: uptime
    - name: Send the uptime to rocketchat channel
      rocketchat:
        token: TOKEN
        domain: chat.example.com
        channel: '#ansible'
        msg: 'Local system uptime is {{ uptime.stdout }}.'
```

As you can see, the only lines that changed are the sixth and seventh, where the word `slack` has been replaced by `rocketchat`. Also, we need to add the domain field specifying where our installation of Rocket Chat is located.

Run the code with the following command:

```
ansible-playbook -i localhost, uptime_and_rocketchat.yaml
```

This results in the following output:

```
PLAY [localhost] ***********************************************
TASK [setup] **************************************************
ok: [localhost]
TASK [Read the machine uptime] ********************************
changed: [localhost]
TASK [Send the uptime to rocketchat channel] *****************
changed: [localhost]
PLAY RECAP ***************************************************
localhost              : ok=3    changed=2    unreachable=0    failed=0
```

Another way to self-host a company's conversations is by using IRC, a very old but still commonly used protocol. Ansible is also able to send messages using it.

Internet Relay Chat

Internet Relay Chat (IRC) is probably the most well-known and widely-used chat protocol of the 1990s and it's still used today. Its popularity and continued use is mainly due to its use in open source communities and its simplicity. From an Ansible perspective, IRC is a pretty straightforward module and we can use it, as shown in the following example (to be put in the uptime_and_irc.yaml file):

```
---
- hosts: localhost
  connection: local
  tasks:
    - name: Read the machine uptime
      command: 'uptime -p'
      register: uptime
    - name: Send the uptime to IRC channel
      irc:
        port: 6669
        server: irc.example.net
        channel: '#desired_channel'
        msg: 'Local system uptime is {{ uptime.stdout }}.'
        color: green
```

 You need the socket Python library installed to use the Ansible IRC module.

In the IRC module, the following fields are required:

- channel: This is to specify in which channel your message will be delivered.
- msg: This is the message you want to send.

Other configurations you will usually specify are:

- server: Select server to connect to, if not localhost.
- port: Select port to connect to, if not 6667.
- color: This to specify the message color, if not black.

- `nick`: This to specify the `nick` sending the message, if not `ansible`.
- `use_ssl`: Use SSL and TLS security.
- `style`: This is if you want to send your message with bold, italic, underline, or reverse style.

Run the code with the following command:

```
ansible-playbook uptime_and_irc.yaml
```

This results in the following output:

```
PLAY [localhost] **********************************************
TASK [setup] **************************************************
ok: [localhost]
TASK [Read the machine uptime] ********************************
changed: [localhost]
TASK [Send the uptime to IRC channel] *************************
changed: [localhost]
PLAY RECAP ****************************************************
localhost         : ok=3    changed=2   unreachable=0    failed=0
```

We have seen many different communication systems that are probably in place in your company or project, but those are usually used for a human-to-human or a machine-to-human communication. Machine-to-machine communication usually uses different systems, such as Amazon SNS.

Amazon Simple Notification Service

Sometimes, you want your playbooks to be agnostic in the way you receive the alert. This has several advantages, mainly in terms of flexibility. In fact, in this model, Ansible will deliver the messages to a notification service and the notification service will then take care of delivering them. **Amazon Simple Notification Service** (**SNS**) is not the only notification service available, but it's probably the most used. SNS has the following components:

- **Messages**: Messages generated by publishers identified by a UUID
- **Publishers**: Programs generating messages
- **Topics**: Named groups of messages, which can be thought of in a similar way to chat channels or rooms
- **Subscribers**: Clients that will receive all messages published in the topics they have subscribed to

So in our case, we will have, specifically, the following:

- **Messages**: Ansible notifications
- **Publishers**: Ansible itself
- **Topics**: Probably different topics to group messages based on the system and/or the kind of notification (for example, storage, networking, or computing)
- **Subscribers**: The people in your team that have to be notified

As we said, one of the big advantages of SNS is that you can decouple the way Ansible sends messages (the SNS API) from the way your users will receive them. In fact, you will be able to choose different delivery systems per user and per topic rules, and eventually you can change them dynamically to ensure that the messages are sent in the best way possible for any situation. The five ways SNS can send messages, at the moment, are as follows:

- Amazon **Lambda** functions (serverless functions written in Python, Java, and JavaScript)
- Amazon **Simple Queue Service** (**SQS**) (a message queuing system)
- Email
- HTTP(S) call
- SMS

Let's see how we can send SNS messages with Ansible. To do so, we can create a file called `uptime_and_sns.yaml` with the following content:

```
---
- hosts: localhost
  connection: local
  tasks:
    - name: Read the machine uptime
      command: 'uptime -p'
      register: uptime
    - name: Send the uptime to SNS
      sns:
        msg: 'Local system uptime is {{ uptime.stdout }}.'
        subject: "System uptime"
        topic: "uptime"
```

In this example, we are using the `msg` key to set the message that will be sent, `topic` to choose the most appropriate topic, and `subject` will be used as the subject for email deliveries. There are many other options you can set. Mainly, they are useful for sending different messages using different delivery methods.

For instance, it would make sense to send a short message via SMS (in the end, the first **S** in **SMS** means **short**) and longer and more detailed messages via emails. To do so, the SNS module provides us with the following delivery-specific options:

- `email`
- `http`
- `https`
- `sms`
- `sqs`

As we have seen in previous chapter, AWS modules require credentials, and we can set them up in multiple ways. The three AWS-specific parameters that are needed to run this module are:

- `aws_access_key`: This is the AWS access key; if not specified, the environmental variable, `aws_access_key`, will be considered or the content of `~/.aws/credentials`.
- `aws_secret_key`: This is the AWS secret key; if not specified, the environmental variable, `aws_secret_key`, will be considered or the content of `~/.aws/credentials`.
- `region`: This is the AWS region to use; if not specified, the environmental variable, `ec2_region`, will be considered or the content of `~/.aws/config`.

Run the code with the following command:

```
ansible-playbook uptime_and_sns.yaml
```

This will result in the following output:

```
PLAY [localhost] ************************************************

TASK [setup] ***************************************************
ok: [localhost]

TASK [Read the machine uptime] *********************************
changed: [localhost]

TASK [Send the uptime to SNS] **********************************
changed: [localhost]

PLAY RECAP ****************************************************
localhost              : ok=3    changed=2    unreachable=0    failed=0
```

There are cases where we want to notify a monitoring system so that it will not trigger any alarm due to Ansible actions. A common example of such a system is Nagios.

Nagios

Nagios is one of the most used tools for controlling the status of services and servers. Nagios is capable of regularly auditing the state of servers and services and notifying users in case of problems. If you have Nagios in your environment, you need to be very careful when you administer your machines, because in cases where Nagios finds servers or services in an unhealthy state, it will start sending emails and SMS and making calls to your team. When you run Ansible scripts against nodes that are controlled by Nagios, you have to be even more careful, because you risk emails, SMS messages, and calls being triggered during the night or other inappropriate times. To avoid this, Ansible is able to notify Nagios beforehand, so that Nagios does not send notifications in that time window, even if some services are down (for instance, because they are rebooted) or other checks fail.

In this example, we are going to stop a service, wait for five minutes, and then start it again since this would actually create a Nagios failure in the majority of configurations. In fact, usually, Nagios is configured to accept up to two consecutive failures of a test (with usually one execution every minute), putting the service in a warning state before raising a critical state. We are going to create the `long_restart_service.yaml` file, which will trigger the Nagios critical state:

```
---
- hosts: ws01.fale.io
  tasks:
    - name: Stop the HTTPd service
      service:
        name: httpd
        state: stopped
    - name: Wait for 5 minutes
      pause:
        minutes: 5
    - name: Start the HTTPd service
      service:
        name: httpd
        state: stopped
```

Run the code with the following command:

```
ansible-playbook long_restart_service.yaml
```

This should trigger a Nagios alert and result in the following output:

```
PLAY [ws01.fale.io] **********************************************

TASK [setup] *****************************************************
ok: [ws01.fale.io]

TASK [Stop the HTTpd service] ************************************
changed: [ws01.fale.io]

TASK [Wait for 5 minutes] ****************************************
changed: [ws01.fale.io]

TASK [Start the HTTpd service] ***********************************
changed: [ws01.fale.io]

PLAY RECAP *******************************************************
ws01.fale.io       : ok=4    changed=3    unreachable=0    failed=0
```

> If no Nagios alert has been triggered, either your Nagios installation does not track that service or five minutes is not enough to make it raise a critical state. To check, you should get in contact to the person or the team that manages your Nagios installation, since Nagios allows full configuration to a point where it's very hard to predict how Nagios will behave without knowing its configuration.

We can now create a very similar playbook that will ensure that Nagios will not send any alerts. We are going to create a file called `long_restart_service_no_alert.yaml` with the following content (the full code is available on GitHub):

```
---
- hosts: ws01.fale.io
  tasks:
    - name: Mute Nagios
      nagios:
        action: disable_alerts
        service: httpd
        host: '{{ inventory_hostname }}'
      delegate_to: nagios.fale.io
    - name: Stop the HTTPd service
      service:
        name: httpd
        state: stopped
...
```

As you can see, we have added two tasks. The first is to inform Nagios not to send alerts for the HTTPd service on the given host, and the second is to inform Nagios to start sending alerts for the service again. Even if you do not specify the service and therefore all alerts on that host are silenced, my advice is to disable only the alert you are going to break so that Nagios is still able to work normally on the majority of your infrastructure.

 If the playbook run fails before reaching the re-enablement of the alerts, your alerts will stay *disabled*.

This module's goal is to toggle the Nagios alerts as well as schedule downtime, and from Ansible 2.2, this module can also unschedule downtimes.

Run the code with the following command:

```
ansible-playbook long_restart_service_no_alert.yaml
```

This should trigger a Nagios alert and result in the following output (the full code output is available on GitHub):

```
PLAY [ws01.fale.io] *******************************************
TASK [setup] **************************************************
ok: [ws01.fale.io]
TASK [Mute Nagios] ********************************************
changed: [nagios.fale.io]
TASK [Stop the HTTpd service] ********************************
changed: [ws01.fale.io]
...
```

 To use the Nagios module, you need to delegate the action to your Nagios server using the delegate_to parameter, as shown in the example.

Sometimes, what you want to achieve with a Nagios integration is exactly the opposite. In fact, you are not interested in muting it, but you want Nagios to handle your test results. A common case is if you want to leverage your Nagios configuration to notify your administrators of the output of a task. To do so, we can use the Nagios nsca utility, integrating it into our playbooks. Ansible does not yet have a specific module for managing it, but you can always run it using the command module, leveraging the send_nsca CLI program.

Summary

In this chapter, we have seen how we can teach Ansible to send notifications to other systems and people. You learned to send notifications through a variety of systems, including email and messaging services, such as Slack. Finally, you learned how to prevent Nagios from sending unwanted notifications about system health during periods when you are operating it.

In the next chapter, we will learn how to create a module so that you can extend Ansible to perform any kind of task.

Section 3: Deploying an Application with Ansible

3

This section explains how to debug and test Ansible to ensure that your playbooks will always work. You will also learn how to manage multiple tiers, multiple environments, and multiple deployments with Ansible.

This section contains the following chapters:

- Chapter 7, *Creating a Custom Module*
- Chapter 8, *Debugging and Error Handling*
- Chapter 9, *Complex Environments*

Creating a Custom Module

7

This chapter will focus on how to write and test custom modules. We've already discussed how modules work and how to use them within your tasks. To quickly recap, a module in Ansible is a piece of code that is transferred and executed onto your remote host every time you run an Ansible task (it can also run locally if you've used `local_action`).

From my experience, I've seen custom modules being written whenever a certain functionality needs to be exposed as a first-class task. The same functionality can be achieved without the module, but it will require a series of tasks with existing modules to accomplish the end goal (and sometimes, command and shell modules too). For example, let's say that you want to provision a server through **Preboot Execution Environment** (**PXE**). Without a custom module, you would probably use a few shell or command tasks to accomplish this. However, with a custom module, you can just pass the required parameters to it and the business logic will be embedded within the custom module in order to perform the PXE boot. This gives you the ability to write playbooks that are much simpler to read, and gives your code better reusability, since you create the module once and can use it everywhere in your roles and playbooks.

The arguments that you pass to a module, provided they are in a **key-value** format, will be forwarded in a separate file along with the module. Ansible expects at least two variables in your module output (that is, the result of the module running), whether it passed or failed, and a message for the user – and they both have to be in JSON format. If you adhere to this simple rule, you can customize your module as much as you want!

In this chapter, we will cover the following topics:

- Python modules
- Bash modules
- Ruby modules
- Testing modules

Prerequisite

When you choose a particular technology or tool, you generally start with what it offers. You slowly understand the philosophy then went into building the tool and what problems it helps you solve. However, you only truly feel comfortable and in control when you understand how it works in depth. At some stage, to utilize the complete power of a tool, you'll have to customize it in ways that suit your particular needs. Over a period of time, tools that provide you with an easy way to plug in new functionalities stay, and those that don't, disappear from the market. It's a similar story with Ansible as well. All tasks in the Ansible playbook are modules of some kind, and it comes loaded with hundreds of modules. You will find a module for almost anything that you might need. However, there are always exceptions, and this is where the power to extend Ansible's functionality, by adding custom modules comes in.

Chef provides **Lightweight Resources and Providers** (**LWRPs**) to perform this activity and Ansible allows you to extend its functionality using custom modules. With Ansible, you can write the module in any language of your choice (provided that you have an interpreter of that language), whereas, in Chef, the module has to be in Ruby. Ansible developers recommend using Python for any complex module, as there is out-of-the-box support to parse arguments; almost all ***nix** systems have Python installed by default and Ansible itself is written in Python. In this chapter, we will also learn how to write modules in other languages.

To make your custom modules available to Ansible, you can do one of the following:

- Specify the path to your custom module in the `ANSIBLE_LIBRARY` environment variable.
- Use the `--module-path` command-line option.
- Drop the modules in the `library` directory in your Ansible top-level directory and add `library=library` in the `[default]` section of your `ansible.cfg` file.

You can download all the files from this book's GitHub repository at `https://github.com/PacktPublishing/Learning-Ansible-2.X-Third-Edition/tree/master/Chapter07`.

Now that we have this background information, let's take a look at some code!

Using Python to write modules

Ansible allows users to write modules in any language. Writing the module in Python, however, has its own advantages. You can take advantage of Ansible's libraries to shorten your code – an advantage that is not available for modules written in other languages. Parsing user arguments, handling errors, and returning the required values becomes easier with the help of the Ansible libraries. Additionally, since Ansible is written in Python, you will have the same language for your whole Ansible code base, making reviews easier and maintainability higher. We will see two examples of a custom Python module, one using the Ansible library and one without, in order to give you a glimpse of how custom modules work. Make sure that you organize your directory structure as mentioned in the previous section before creating the module. The first example creates a module named check_user. To do this, we will need to create the check_user.py file in the library folder within the Ansible top-level directory. The full code is available on GitHub:

```python
def main():
    # Parsing argument file
    args = {}
    args_file = sys.argv[1]
    args_data = file(args_file).read()
    arguments = shlex.split(args_data)
    for arg in arguments:
        if '=' in arg:
            (key, value) = arg.split('=')
            args[key] = value
    user = args['user']

    # Check if user exists
    try:
        pwd.getpwnam(user)
        success = True
        ret_msg = 'User %s exists' % user
    except KeyError:
        success = False
        ret_msg = 'User %s does not exists' % user
    ...
```

The preceding custom module, `check_user`, will check whether a user exists on a host. The module expects a user argument from Ansible. Let's break down the preceding module and see what it does. We first declare the Python interpreter and import the libraries that are required to parse the arguments:

```
#!/usr/bin/env python

import pwd
import sys
import shlex
import json
```

Using the `sys` library, we then parse the arguments, which are passed in a file by Ansible. The arguments are in the `param1=value1 param2=value2` format, where `param1` and `param2` are the parameters, and `value1` and `value2` are the values of the parameters. There are multiple ways in which to split arguments and create a dictionary, and we've chosen an easy way to perform the operation. We first create a list of arguments by splitting the arguments with a whitespace character, and then we separate the key and value by splitting the arguments with an = character and assigning it to a Python dictionary. For example, if you have a string such as `user=foo gid=1000`, then you will first create a list, `["user=foo", "gid=1000"]`, and then loop over this list to create a dictionary. This dictionary will be `{"user": "foo", "gid": 1000}`; this is performed using the following lines:

```
def main():
    # Parsing argument file
    args = {}
    args_file = sys.argv[1]
    args_data = file(args_file).read()
    arguments = shlex.split(args_data)
    for arg in arguments:
        if '=' in arg:
            (key, value) = arg.split('=')
            args[key] = value
    user = args['user']
```

 We separate the arguments based on a whitespace character, because this is the standard, followed by core Ansible modules. You can use any separator instead of whitespace, but we encourage you to maintain uniformity.

Once we have the user argument, we then check whether this user exists on the host, as follows:

```
# Check if user exists
try:
    pwd.getpwnam(user)
    success = True
    ret_msg = 'User %s exists' % user
except KeyError:
    success = False
    ret_msg = 'User %s does not exists' % user
```

We use the `pwd` library to check the `passwd` file for the user. For the sake of simplicity, we use two variables: one to store the success or failure message and the other to store the message for the user. Finally, we use the variables created in the `try-catch` block to check whether the module succeeded or failed:

```
# Error handling and JSON return
if success:
    print json.dumps({
        'msg': ret_msg
    })
    sys.exit(0)
else:
    print json.dumps({
        'failed': True,
        'msg': ret_msg
    })
    sys.exit(1)
```

If the module succeeds, then it will exit the execution with an exit code of 0 (`exit(0)`); otherwise, it will exit with a non-zero code. Ansible will look for the failed variable and, if it is set to `True`, it will exit unless you have explicitly asked Ansible to ignore the error using the `ignore_errors` parameter. You can use customized modules like any other core module of Ansible. To test the custom module, we will need a playbook, so let's create the `playbooks/check_user.yaml` file using the following code:

```
---
- hosts: localhost
  connection: local
  vars:
    user_ok: root
    user_ko: this_user_does_not_exists
  tasks:
    - name: 'Check if user {{ user_ok }} exists'
      check_user:
```

```
        user: '{{ user_ok }}'
    - name: 'Check if user {{ user_ko }} exists'
      check_user:
        user: '{{ user_ko }}'
```

As you can see, we used the `check_user` module like any other core module. Ansible will execute this module onto the remote host by copying the module to the remote host with the arguments in a separate file. Let's see how this playbook runs with the following code:

```
ansible-playbook playbooks/check_user.yaml
```

We should receive the following output:

```
PLAY [localhost] **************************************************

TASK [Gathering Facts] *******************************************
ok: [localhost]

TASK [Check if user root exists] *********************************
ok: [localhost]

TASK [Check if user this_user_does_not_exists exists] ***********
fatal: [localhost]: FAILED! => {"changed": false, "msg": "User
this_user_does_not_exists does not exists"}
        to retry, use: --limit @playbooks/check_user.retry

PLAY RECAP *****************************************************
localhost                  : ok=2 changed=0 unreachable=0 failed=1
```

As expected, since we have the `root` user, but not `this_user_does_not_exists`, it passed the first check, but failed at the second.

Ansible also provides a Python library to parse user arguments and handle errors and returns. It's time to explore how the Ansible Python library can be used to make your code shorter, faster, and less prone to error. To do so, let's create a file called `library/check_user_py2.py` with the following code. The full code is available on GitHub:

```python
#!/usr/bin/env python

import pwd
from ansible.module_utils.basic import AnsibleModule

def main():
    # Parsing argument file
    module = AnsibleModule(
        argument_spec = dict(
```

```
                user = dict(required=True)
            )
        )
        user = module.params.get('user')

        # Check if user exists
        try:
            pwd.getpwnam(user)
            success = True
            ret_msg = 'User %s exists' % user
        except KeyError:
            success = False
            ret_msg = 'User %s does not exists' % user

    ...
```

Let's break down the preceding module and see how it works, as follows:

```
#!/usr/bin/env python

import pwd
from ansible.module_utils.basic import AnsibleModule
```

As you can see, we do not import `sys`, `shlex`, and `json`; we are no longer using them since all the operations that required them are now done by the Ansible `module_utils` module:

```
    # Parsing argument file
    module = AnsibleModule(
        argument_spec = dict(
            user = dict(required=True)
        )
    )
    user = module.params.get('user')
```

Previously, we performed a lot of processing on the argument file to get the final user arguments. Ansible makes it easy by providing an `AnsibleModule` class, which does all the processing on its own and provides us with the final arguments. The `required=True` parameter means that the argument is mandatory and the execution will fail if the argument is not passed. The default value required is `False`, which will allow users to skip the argument. You can then access the value of the arguments through the `module.params` dictionary by calling the `get` method on `module.params`. The logic to check users on the remote host will remain the same, but the error handling and return aspect will change as follows:

```
    # Error handling and JSON return
    if success:
        module.exit_json(msg=ret_msg)
```

```
        else:
            module.fail_json(msg=ret_msg)
```

One of the advantages of using the `AnsibleModule` object is that you have a very nice facility to handle returning values to the playbook. We will go into more depth in the next section.

 We could have condensed the logic for checking users and the return aspect, but we kept them divided for readability.

To verify that everything works as expected, we can create a new playbook in `playbooks/check_user_py2.yaml` with the following code:

```
---
- hosts: localhost
  connection: local
  vars:
    user_ok: root
    user_ko: this_user_does_not_exists
  tasks:
    - name: 'Check if user {{ user_ok }} exists'
      check_user_py2:
        user: '{{ user_ok }}'
    - name: 'Check if user {{ user_ko }} exists'
      check_user_py2:
        user: '{{ user_ko }}'
```

You can run it with the following code:

```
ansible-playbook playbooks/check_user_py2.yaml
```

Then, we should receive the following output:

```
PLAY [localhost] *********************************************

TASK [Gathering Facts] ***************************************
ok: [localhost]

TASK [Check if user root exists] *****************************
ok: [localhost]

TASK [Check if user this_user_does_not_exists exists] ********
fatal: [localhost]: FAILED! => {"changed": false, "msg": "User
this_user_does_not_exists does not exists"}
        to retry, use: --limit @playbooks/check_user_py2.retry
```

```
PLAY RECAP *********************************************************
localhost                   : ok=2 changed=0 unreachable=0 failed=1
```

This output is consistent with our expectations.

Working with exit_json and fail_json

Ansible provides a faster and shorter method for handling success and failure through the `exit_json` and `fail_json` methods, respectively. You can directly pass a message to these methods and Ansible will take care of the rest. You can also pass additional variables to these methods and Ansible will print those variables to `stdout`. For example, apart from the message, you might also want to print the `uid` and `gid` parameters of the user. You can do this by passing these variables to the `exit_json` method, separated by a comma.

Let's take a look at how you can return multiple values to `stdout`, as demonstrated in the following code, placed in `library/check_user_id.py`. The full code is available on GitHub:

```python
#!/usr/bin/env python

import pwd
from ansible.module_utils.basic import AnsibleModule

class CheckUser:
    def __init__(self, user):
        self.user = user

    # Check if user exists
    def check_user(self):
        uid = ''
        gid = ''
        try:
            user = pwd.getpwnam(self.user)
            success = True
            ret_msg = 'User %s exists' % self.user
            uid = user.pw_uid
            gid = user.pw_gid
        except KeyError:
            success = False
            ret_msg = 'User %s does not exists' % self.user
        return success, ret_msg, uid, gid

...
```

As you can see, we return the `uid` and `gid` parameters of the user, along with the message, `msg`. You can have multiple values and Ansible will print all of them in dictionary format. Create a playbook in `playbooks/check_user_id.yaml` with the following content:

```
---
- hosts: localhost
  connection: local
  vars:
    user: root
  tasks:
    - name: 'Retrive {{ user }} data if it exists'
      check_user_id:
        user: '{{ user }}'
      register: user_data
    - name: 'Print user {{ user }} data'
      debug:
        msg: '{{ user_data }}'
```

You can run it with the following code:

```
ansible-playbook playbooks/check_user_id.yaml
```

We should receive the following output:

```
PLAY [localhost] **************************************************

TASK [Gathering Facts] *******************************************
ok: [localhost]

TASK [Retrive root data if it exists] **************************
ok: [localhost]

TASK [Print user root data] ************************************
ok: [localhost] => {
 "msg": {
 "changed": false,
 "failed": false,
 "gid": 0,
 "msg": "User root exists",
 "uid": 0
 }
}

PLAY RECAP ****************************************************
localhost : ok=3 changed=0 unreachable=0 failed=0
```

Here, we completed the working of both methods, which, in turn, helped us to find a quicker way to handle success and failure in Ansible while passing parameters to the user.

Testing Python modules

As you have seen, you can test your modules by creating very simple playbooks to run them. To do so, we'll need to clone the Ansible official repository (if you haven't done so yet):

```
git clone git://github.com/ansible/ansible.git --recursive
```

Next, source an environmental file, as follows:

```
source ansible/hacking/env-setup
```

We can now use the `test-module` utility to run the script by passing the filename as a command-line argument:

```
ansible/hacking/test-module -m library/check_user_id.py -a "user=root"
```

The result will be similar to the following output:

```
* including generated source, if any, saving to:
/home/fale/.ansible_module_generated
* ansiballz module detected; extracted module source to:
/home/fale/debug_dir
***********************************
RAW OUTPUT

{"msg": "User root exists", "invocation": {"module_args": {"user":
"root"}}, "gid": 0, "uid": 0, "changed": false}


***********************************
PARSED OUTPUT
{
    "changed": false,
    "gid": 0,
    "invocation": {
        "module_args": {
            "user": "root"
        }
    },
    "msg": "User root exists",
    "uid": 0
}
```

It's also easy to execute the script directly if you have not used `AnsibleModule`. This is because this module requires a lot of Ansible-specific variables, so it's more complicated to "simulate" an Ansible run than to actually run Ansible itself.

Using bash modules

Bash modules in Ansible are no different to any other bash scripts, except in the way that they print the data on `stdout`. Bash modules can be as straightforward as checking whether a process is running on the remote host, to running some more complex commands.

As previously stated, the general recommendation is to use Python for modules. In my opinion, the second-best choice (only for very easy modules) is the `bash` module, due to its simplicity and user base.

Let's create a `library/kill_java.sh` file with the following content:

```bash
#!/bin/bash
source $1

SERVICE=$service_name

JAVA_PIDS=$(/usr/java/default/bin/jps | grep ${SERVICE} | awk '{print $1}')

if [ ${JAVA_PIDS} ]; then
    for JAVA_PID in ${JAVA_PIDS}; do
        /usr/bin/kill -9 ${JAVA_PID}
    done
    echo "failed=False msg=\"Killed all the orphaned processes for
${SERVICE}\""
    exit 0
else
    echo "failed=False msg=\"No orphaned processes to kill for
${SERVICE}\""
    exit 0
fi
```

The preceding `bash` module will take the `service_name` argument and forcefully kill all of the Java processes that belong to that service. As you know, Ansible passes the argument file to the module. We then source the argument file using the `$1` source. This will actually set the environment variable with the name, `service_name`. We then access this variable using `$service_name`, as follows:

```
source $1

SERVICE=$service_name
```

We then check to see if we obtained any `PIDS` for the service and run a loop over it to forcefully kill all of the Java processes that match `service_name`. Once they're killed, we exit the module with `failed=False` and a message with an exit code of `0`, as you can see in the following code:

```
if [ ${JAVA_PIDS} ]; then
    for JAVA_PID in ${JAVA_PIDS}; do
        /usr/bin/kill -9 ${JAVA_PID}
    done
    echo "failed=False msg=\"Killed all the orphaned processes for
${SERVICE}\""
    exit 0
```

If we do not find any running processes for the service, we will still exit the module with an exit code of `0`, because terminating the Ansible run might not make sense:

```
else
    echo "failed=False msg=\"No orphaned processes to kill for
${SERVICE}\""
    exit 0
fi
```

 You can also terminate the Ansible run by printing `failed=True` with an exit code of `1`.

Ansible allows you to return a key-value output if the language itself doesn't support JSON. This makes Ansible more developer and sysadmin-friendly, and allows custom modules to be written in any language of your choice. Let's test the `bash` module by passing the arguments file to the module. We can now create an arguments file in `/tmp/arguments` that has the `service_name` parameter set to `jenkins`, as follows:

```
service_name=jenkins
```

Now, you can run the module like any other bash script. Let's see what happens when we run it with the following code:

```
bash library/kill_java.sh /tmp/arguments
```

We should receive the following output:

```
failed=False msg="No orphaned processes to kill for jenkins"
```

As expected, the module did not fail even though there was no Jenkins process running on the localhost.

If you receive the `jps command does not exists` error instead of the mentioned output, your machine is probably missing Java. If so, you can install it by following the instructions for your operating system at `https://www.java.com/en/download/help/download_options.xml`.

Using Ruby modules

Writing modules in Ruby is as easy as writing a module in Python or bash. You just need to take care of the arguments, errors, return statements, and, of course, know basic Ruby! Let's create the `library/rsync.rb` file with the following code. The full code is available on GitHub:

```
#!/usr/bin/env ruby

require 'rsync'
require 'json'

src = ''
dest = ''
ret_msg = ''
SUCCESS = ''

def print_message(state, msg, key='failed')
    message = {
```

```
        key => state,
        "msg" => msg
    }
    print message.to_json
    exit 1 if state == false
    exit 0
...
```

In the preceding module, we first process the user arguments, then copy the file using the `rsync` library, and finally, return the output.

To be able to use this, you need to ensure that the `rsync` library for Ruby is present on your system. To do so, you can execute the following command:

gem install rsync

Let's break down the preceding code and see how it works.

We first write a method, `print_message`, which will print the output in a JSON format. By doing this, we can reuse the same code in multiple places. Remember, the output of your module should contain `failed=true` if you want the Ansible run to fail; otherwise, Ansible will think that the module succeeded and will continue with the next task. The output that is obtained is as follows:

```ruby
#!/usr/bin/env ruby

require 'rsync'
require 'json'

src = ''
dest = ''
ret_msg = ''
SUCCESS = ''

def print_message(state, msg, key='failed')
    message = {
        key => state,
        "msg" => msg
    }
    print message.to_json
    exit 1 if state == false
    exit 0
end
```

We then process the argument file, which contains a key-value pair separated by a whitespace character. This is similar to what we did with the Python module earlier, where we took care of parsing out the arguments. We also perform some checks to make sure that the user has not missed any required argument. In this case, we check if the `src` and `dest` parameters have been specified, and print a message if the arguments are not provided. Further checks could include the format and type of arguments. You can add these checks and any other checks that you deem important. For example, if one of your parameters is a `date` parameter, then you'd need to verify that the input is actually the right date. Consider the following piece of code, which shows the discussed parameters:

```
args_file = ARGV[0]
data = File.read(args_file)
arguments = data.split(" ")
arguments.each do |argument|
    print_message(false, "Argument should be name-value pairs. Example
name=foo") if not argument.include?("=")
    field, value = argument.split("=")
    if field == "src"
        src = value
    elsif field == "dest"
        dest = value
    else print_message(false, "Invalid argument provided. Valid arguments
are src and dest.")
    end
end
```

Once we have the required arguments, we will go ahead and copy the file using the `rsync` library, as follows:

```
result = Rsync.run("#{src}", "#{dest}")
if result.success?
    success = true
    ret_msg = "Copied file successfully"
else
    success = false
    ret_msg = result.error
end
```

Finally, we check if the `rsync` task passed or failed, and then call the `print_message` function to print the output on `stdout`, as follows:

```
if success
    print_message(false, "#{ret_msg}")
else
    print_message(true, "#{ret_msg}")
end
```

You can test your Ruby module by simply passing the arguments file to the module. To do so, we can create the /tmp/arguments file with the following content:

```
src=/etc/resolv.conf dest=/tmp/resolv_backup.conf
```

Let's now run the module, as follows:

```
ruby library/rsync.rb /tmp/arguments
```

We will receive the following output:

```
{"failed":false,"msg":"Copied file successfully"}
```

We will leave the serverspec testing for you to complete.

Testing modules

Testing is often undervalued because of the lack of understanding of its purpose and the benefits that it can bring to a business. Testing modules are as important as testing any other part of the Ansible playbook, because a small change in a module can break your entire playbook. We will take the example of the Python module that we wrote in the *Using Python to write modules* section of this chapter and write an integration test using Python's nose test framework. Unit tests are also encouraged, but for our scenario, where we check whether a user exists remotely, an integration test makes more sense.

 nose is a Python test framework; you can find out more information about this test framework at https://nose.readthedocs.org/en/latest/.

To test the module, we convert our previous module into a Python class so that we can directly import the class into our test, and run only the main logic of the module. The following code shows the library/check_user_py3.py restructured module, which will check whether a user exists on a remote host. The full code is available on GitHub:

```
#!/usr/bin/env python

import pwd
from ansible.module_utils.basic import AnsibleModule

class User:
    def __init__(self, user):
        self.user = user
```

```
# Check if user exists
def check_if_user_exists(self):
    try:
        user = pwd.getpwnam(self.user)
        success = True
        ret_msg = 'User %s exists' % self.user
    except KeyError:
        success = False
        ret_msg = 'User %s does not exists' % self.user
    return success, ret_msg

. . .
```

As you can see in the preceding code, we created a class named `User`. We instantiated the class and called the `check_if_user_exists` method to check whether the user actually exists on the remote machine. It's time to write an integration test now. We assume that you have the `nose` package installed on your system. If not, don't worry! You can still install the package using the following command:

```
pip install nose
```

Let's now write the integration test file in `library/test_check_user_py3.py`, as follows:

```
from nose.tools import assert_equals, assert_false, assert_true
import imp
imp.load_source("check_user","check_user_py3.py")
from check_user import User

def test_check_user_positive():
    chkusr = User("root")
    success, ret_msg = chkusr.check_if_user_exists()
    assert_true(success)
    assert_equals('User root exists', ret_msg)

def test_check_user_negative():
    chkusr = User("this_user_does_not_exists")
    success, ret_msg = chkusr.check_if_user_exists()
    assert_false(success)
    assert_equals('User this_user_does_not_exists does not exists',
ret_msg)
```

In the preceding integration test, we import the `nose` package and our `check_user` module. We call the `User` class by passing the user that we want to check. We then check whether the user exists on the remote host by calling the `check_if_user_exists()` method. The `nose` methods – `assert_true`, `assert_false`, and `assert_equals` – can be used to compare the expected value against the actual value. Only if the `assert` methods pass, will the test also pass. You can have multiple tests inside the same file by having multiple methods whose names start with `test_`; for example, the `test_check_user_positive()` and `test_check_user_negative()` methods. `nose` tests will take all the methods that start with `test_` and execute them.

> As you can see, we actually created two tests for just one function. This is a key part of tests. Always try cases where you know it will work, but also, do not forget to test cases where you expect it to fail.

We can now test whether it works by running `nose` with the following code:

```
cd library
nosetests -v test_check_users_py3.py
```

You should receive an output that is similar to the following code block:

```
test_check_user_py3.test_check_user_positive ... ok
test_check_user_py3.test_check_user_negative ... ok
----------------------------------------------------
Ran 2 tests in 0.001sOK
```

As you can see, the test passed because the `root` user existed on the host, while the `this_user_does_not_exists` user did not.

> We use the `-v` option with `nose` tests for the **verbose** mode.

For more complicated modules, we recommend that you write unit tests and integration tests. You might be wondering why we didn't use `serverspec` to test the module.

We still recommend running `serverspec` tests for functional testing as part of playbooks; however, for unit and integration tests, it's recommended to use well-known frameworks. Similarly, if you write Ruby modules, we recommend that you write tests for them with a framework such as `rspec`. If your custom Ansible module has multiple parameters with multiple combinations, then you will write more tests to test each scenario. Finally, we recommend that you run all these tests as part of your CI system, be it Jenkins, Travis, or any other system.

Summary

With this, we come to the end of this rather small but important chapter, which focused on how you can extend Ansible by writing your own custom modules. You learned how to use Python, bash, and Ruby in order to write your modules. We've also learned how to write integration tests for modules so that they can be integrated into your CI system. In the future, hopefully, extending your Ansible functionality by using modules will be much easier!

In the next chapter, we will step into the world of provisioning, deployment, and orchestration, and look at how Ansible solves our infrastructure problems when we provision new instances or want to deploy software updates to various instances in our environments. We promise that this journey will be fun!

8
Debugging and Error Handling

Like software code, testing infrastructure code is an all-important task. There should ideally be no code floating around in production that has not been tested, especially when you have strict customer SLAs to meet, and this is true even for the infrastructure. In this chapter, we'll look at syntactic checks, testing without applying the code on the machines (the no-op mode), and functional testing for playbooks, which are at the core of Ansible and trigger the various tasks you want to perform on the remote hosts. It is recommended that you integrate some of these into your **Continuous Integration** (**CI**) system that you have for Ansible to better test your playbooks. We'll be looking at the following points:

- Syntax checking
- Checking mode with and without `--diff`
- Functional testing

As part of functional testing, we will be looking at the following:

- Assertions on the end state of the system
- Testing with tags
- Using the `--syntax-check` option
- Using the `ANSIBLE_KEEP_REMOTE_FILES` and `ANSIBLE_DEBUG` flags

We will then look at how to manage exceptions and how to voluntarily trigger an error.

Technical requirements

For this chapter, there are no specific requirements except the usual ones, such as Ansible, Vagrant, and a shell.

You can download all of the files from this book's GitHub repository at `https://github.com/PacktPublishing/Learning-Ansible-2.X-Third-Edition/tree/master/Chapter08`.

Syntax checking

Whenever you run a playbook, Ansible first checks the syntax of the playbook file. If an error is encountered, Ansible will error out saying there was a syntax error and will not proceed unless you fix that error. This syntax checking is performed only when you run the `ansible-playbook` command. When writing a big playbook, or if you have included task files, it might be difficult to fix all of the errors; this might end up wasting more time. In order to deal with such situations, Ansible provides a way to check your YAML syntax as you keep progressing with your playbook. For this example, we will need to create the `playbooks/setup_apache.yaml` file with the following content:

```
---
- hosts: all
  tasks:
    - name: Install Apache
      yum:
        name: httpd
        state: present
      become: True
    - name: Enable Apache
      service:
        name: httpd
        state: started
        enabled: True
      become: True
```

Now that we have our example file, we need to run it with the `--syntax-check` parameter; so, you will need to invoke Ansible as follows:

```
ansible-playbook playbooks/setup_apache.yaml --syntax-check
```

The `ansible-playbook` command checked the YAML syntax of the `setup_apache.yml` playbook and showed that the syntax of the playbook was correct. Let's look at the resulting errors from the invalid syntax in the playbook:

```
ERROR! Syntax Error while loading YAML.
  did not find expected '-' indicator

The error appears to have been in '/home/fale/Learning-Ansible-2.X-Third-
Edition/Ch8/playbooks/setup_apache.yaml': line 10, column 5, but may
be elsewhere in the file depending on the exact syntax problem.

The offending line appears to be:

   - name: Enable Apache
   service:
   ^ here
```

The error shows that there is an indentation error in the `Enable Apache` task. Ansible also gives you the line number, column number, and the filename where this error is found (even if this is not a guarantee of the exact location of the error). This should definitely be one of the basic tests that you should run as part of your CI for Ansible.

The check mode

The check mode (also known as the **dry-run** or **no-op mode**) will run your playbook in a no-operation mode—that is, it will not apply any changes to the remote host; instead, it will just show the changes that will be introduced when a task is run. Whether the check mode is actually enabled or not depends on each module. There are few commands that you may find interesting. All of these commands will have to be run in `/usr/lib/python2.7/site-packages/ansible/modules`, or where your Ansible module folder is (different paths could be possible based on the operating system you are using as well as the way you installed Ansible).

To count the number of available modules on your installation, you can perform this command:

```
find . -type f | grep '.py$' | grep -v '__init__' | wc -l
```

With Ansible 2.7.2, the result of this command is `2095`, since Ansible has that many modules.

If you want to see how many of these support the check mode, you can run the following code:

```
grep -r 'supports_check_mode=True' | awk -F: '{print $1}' | sort | uniq |
wc -l
```

With Ansible 2.7.2, the result of this command is 1239.

You might also find the following command useful for listing all modules that support the check mode:

```
grep -r 'supports_check_mode=True' | awk -F: '{print $1}' | sort | uniq
```

This helps you to test how your playbook will behave and check whether there may be any failures before running it on your production server. You run a playbook in the check mode by simply passing the --check option to your ansible-playbook command. Let's see how the check mode works with the setup_apache.yml playbook, running the following code:

```
ansible-playbook --check -i ws01, playbooks/setup_apache.yaml
```

The result will be as follows:

```
PLAY [all] ************************************************************

TASK [Gathering Facts] **********************************************
ok: [ws01]

TASK [Install Apache] ***********************************************
changed: [ws01]

TASK [Enable Apache] ************************************************
changed: [ws01]

PLAY RECAP ********************************************************
ws01                        : ok=3 changed=2 unreachable=0 failed=0
```

In the preceding run, instead of making the changes on the target host, Ansible highlighted all of the changes that would have occurred during the actual run. From the preceding run, you can find that the httpd service was already installed on the target host. Because of this, Ansible's exit message for that task was OK:

```
TASK [Install Apache] ***********************************************
changed: [ws01]
```

However, with the second task, it found that the `httpd` service was not running on the target host:

```
TASK [Enable Apache] *********************************************
changed: [ws01]
```

When you run the preceding playbook again without the check mode enabled, Ansible will make sure that the service state is running.

Indicating differences between files using --diff

In the check mode, you can use the `--diff` option to show the changes that would be applied to a file. To be able to see the `--diff` option in use, we need to create a `playbooks/setup_and_config_apache.yaml` playbook to match the following:

```
- hosts: all
  tasks:
    - name: Install Apache
      yum:
        name: httpd
        state: present
      become: True
    - name: Enable Apache
      service:
        name: httpd
        state: started
        enabled: True
      become: True
    - name: Ensure Apache userdirs are properly configured
      template:
        src: ../templates/userdir.conf
        dest: /etc/httpd/conf.d/userdir.conf
      become: True
```

As you can see, we added a task that will ensure a certain state of the `/etc/httpd/conf.d/userdir.conf` file.

We also need to create a template file placed in `templates/userdir.conf` with the following content (the full file is available on GitHub):

```
#
# UserDir: The name of the directory that is appended onto a user's home
# directory if a ~user request is received.
```

```
#
# The path to the end user account 'public_html' directory must be
# accessible to the webserver userid. This usually means that ~userid
# must have permissions of 711, ~userid/public_html must have permissions
# of 755, and documents contained therein must be world-readable.
# Otherwise, the client will only receive a "403 Forbidden" message.
#
<IfModule mod_userdir.c>
    #
    # UserDir is disabled by default since it can confirm the presence
    # of a username on the system (depending on home directory
    # permissions).
    #
    UserDir enabled


    ...
```

In this template, we only changed the `UserDir enabled` line, which, by default, is `UserDir disabled`.

> The `--diff` option doesn't work with the `file` module; you will have to use the `template` module only.

We can now test the result of this with the following command:

```
ansible-playbook -i ws01, playbooks/setup_and_config_apache.yaml --diff --check
```

As you can see, we are using the `--check` parameter that will ensure this will be a dry run. We will receive the following output:

```
PLAY [all] ***********************************************************

TASK [Gathering Facts] **********************************************
ok: [ws01]

TASK [Install Apache] ***********************************************
ok: [ws01]

TASK [Enable Apache] ************************************************
ok: [ws01]

TASK [Ensure Apache userdirs are properly configured] ***************
--- before: /etc/httpd/conf.d/userdir.conf
+++ after: /home/fale/.ansible/tmp/ansible-
```

```
local-6756FTSbL0/tmpx9WVXs/userdir.conf
@@ -14,7 +14,7 @@
    # of a username on the system (depending on home directory
    # permissions).
    #
- UserDir disabled
+ UserDir enabled

    #
    # To enable requests to /~user/ to serve the user's public_html

changed: [ws01]

PLAY RECAP ********************************************************
ws01                        : ok=4 changed=1 unreachable=0 failed=0
```

As we can see, Ansible compares the current file of the remote host with the source file; a line starting with + indicates that a line was added to the file, while – indicates that a line was removed.

 You can also use --diff without the --check option, which will allow Ansible to make the specified changes and show the difference between two files.

Using the --diff and --check modes together is a test step that can potentially be used as part of your CI tests to assert how many steps have changed as part of the run. Another case where you can use those features together is the part of the deployment process that checks what exactly will change when you run Ansible on that machine.

There are also cases—which should not happen, but sometimes do—where you have not run a playbook on a machine for a very long time and you are worried that running it again will break something. Using those options together should help you to understand whether it was just worrying you or whether this is a real risk.

Functional testing in Ansible

Wikipedia says functional testing is a **quality assurance** (**QA**) process and a type of black-box testing that bases its test cases on the specifications of the software component under the test. Functions are tested by feeding them input and examining the output; the internal program structure is rarely considered. Functional testing is as important as code when it comes to infrastructure.

From an infrastructure perspective, with respect to functional testing, we test output of our Ansible runs on the actual machines. Ansible provides multiple ways to perform the functional testing of your playbook; let's look at some of the most commonly used methods.

Functional testing using assert

The check mode will only work when you want to check whether a task will change anything on the host or not. This will not help when you want to check whether the output of your module is what you expected. For example, let's say you wrote a module that will check whether a port is up or not. In order to test this, you might need to check the output of your module and see whether it matches the desired output or not. To perform such tests, Ansible provides a way to directly compare the output of a module with the desired output.

Let's see how this works by creating the `playbooks/assert_ls.yaml` file with the following content:

```
---
- hosts: all
  tasks:
    - name: List files in /tmp
      command: ls /tmp
      register: list_files
    - name: Check if file testfile.txt exists
      assert:
        that:
          - "'testfile.txt' in list_files.stdout_lines"
```

In the preceding playbook, we're running the `ls` command on the target host and registering the output of that command in the `list_files` variable. Further to this, we ask Ansible to check whether the output of the `ls` command has the expected result. We do this using the `assert` module, which uses some conditional checks to verify whether or not the `stdout` value of a task meets the expected output of the user. Let's run the preceding playbook to see what output Ansible returns, with the following command:

```
ansible-playbook -i ws01, playbooks/assert_ls.yaml
```

Since we don't have the file, we will receive the following output:

```
PLAY [all] ************************************************************

TASK [Gathering Facts] **********************************************
ok: [ws01]

TASK [List files in /tmp] *******************************************
changed: [ws01]

TASK [Check if file testfile.txt exists] ***************************
fatal: [ws01]: FAILED! => {
    "assertion": "'testfile.txt' in list_files.stdout_lines",
    "changed": false,
    "evaluated_to": false,
    "msg": "Assertion failed"
}
        to retry, use: --limit @/home/fale/Learning-Ansible-2.X-Third-
Edition/Ch8/playbooks/assert_ls.retry

PLAY RECAP ********************************************************
ws01                        : ok=2 changed=1 unreachable=0 failed=1
```

If we re-run the playbook after we create the expected file, this will be the result:

```
PLAY [all] ************************************************************

TASK [Gathering Facts] **********************************************
ok: [ws01]

TASK [List files in /tmp] *******************************************
changed: [ws01]

TASK [Check if file testfile.txt exists] ***************************
ok: [ws01] => {
    "changed": false,
    "msg": "All assertions passed"
}

PLAY RECAP ********************************************************
ws01                        : ok=3 changed=1 unreachable=0 failed=0
```

This time, the task passed with an OK message as `testfile.txt` was present in the `list_files` variable. Likewise, you can match multiple strings in a variable or multiple variables using the `and` and `or` operators. The assertion feature is quite powerful, and users who have written either unit or integration tests in their projects will be quite happy to see this feature!

Testing with tags

Tags are a great way to test a bunch of tasks without running an entire playbook. We can use tags to run actual tests on the nodes to verify the state that the user intended to be in the playbook. We can treat this as another way to run integration tests for Ansible on the actual box. The tag method to test can be run on the actual machines where you run Ansible, and it can be used primarily during deployments to test the state of your end systems. In this section, we'll first look at how to use `tags` in general, their features that can possibly help us, not just with testing but even for testing purposes.

To add tags in your playbook, use the `tags` parameter followed by one or more tag names separated by commas or YAML lists. Let's create a simple playbook in `playbooks/tags_example.yaml` to see how the tags work with the following content:

```
- hosts: all
  tasks:
    - name: Ensure the file /tmp/ok exists
      file:
        name: /tmp/ok
        state: touch
      tags:
        - file_present
    - name: Ensure the file /tmp/ok does not exists
      file:
        name: /tmp/ok
        state: absent
      tags:
        - file_absent
```

If we now run the playbook, the file will be created and destroyed. We can see it running with the following:

```
ansible-playbook -i ws01, playbooks/tags_example.yaml
```

It will give us this output:

```
PLAY [all] ********************************************************

TASK [Gathering Facts] ********************************************
ok: [ws01]

TASK [Ensure the file /tmp/ok exists] *****************************
changed: [ws01]

TASK [Ensure the file /tmp/ok does not exists] ********************
changed: [ws01]

PLAY RECAP ********************************************************
ws01                          : ok=3 changed=2 unreachable=0 failed=0
```

Since this is not an idempotent playbook, if we run it over and over, we will always see the same result, as the playbook will create and delete the file every time.

However, we added two tags: file_present and file_absent. You can now simply pass the file_present tag or the file_absent tag to only perform one of the actions, like in the following example:

```
ansible-playbook -i ws01, playbooks/tags_example.yaml -t file_present
```

Thanks to the -t file_present part, only the tasks with the file_present tag will be executed; in fact, this will be the output:

```
PLAY [all] ********************************************************

TASK [Gathering Facts] ********************************************
ok: [ws01]

TASK [Ensure the file /tmp/ok exists] *****************************
changed: [ws01]

PLAY RECAP ********************************************************
ws01                          : ok=2 changed=1 unreachable=0 failed=0
```

You can also use tags to perform a set of tasks on the remote host, just like taking a server out of a load balancer and adding it back to the load balancer.

You can also use the --check option with tags. By doing this, you can test your tasks without actually running them on your hosts. This allows you to test a bunch of individual tasks directly, instead of copying your tasks to a temporary playbook and running it from there.

Understanding the --skip-tags option

Ansible also provides a way to skip some tags in a playbook. If you have a long playbook with multiple tags, such as 10, and you want to execute them all but one, then it would not be a good idea to pass nine tags to Ansible. The situation would be more difficult if you forgot to pass a tag and the `ansible-run` command fails. To overcome such situations, Ansible provides a way to skip a couple of tags, instead of passing multiple tags, which should run. Its functioning is pretty straightforward, and can be triggered in the following way:

```
ansible-playbook -i ws01, playbooks/tags_example.yaml --skip-tags
file_present
```

The output will be something like this:

```
PLAY [all] ****************************************************

TASK [Gathering Facts] ***************************************
ok: [ws01]

TASK [Ensure the file /tmp/ok exists] ***********************
changed: [ws01]

PLAY RECAP **************************************************
ws01                      : ok=2 changed=1 unreachable=0 failed=0
```

As you can see, all tasks have been executed except the one with the `file_present` tag.

Understanding debugging commands

Ansible allows us to use two very powerful variables to help us to debug.

The `ANSIBLE_KEEP_REMOTE_FILES` variable allows us to tell Ansible to keep the files it creates on remote machines so that we can go back and debug them.

The `ANSIBLE_DEBUG` variable allows us to tell Ansible to print all debug content to the shell. The debug output is often overkill, but it might help with some very complex-to-solve issues.

We have seen how to find problems in your playbooks. Sometimes, you know that a specific step can fail, but it's OK. In this case, we should manage the exception properly. Let's look at how to do so.

Managing exceptions

There are many cases where, for one reason or another, you want your playbook and roles to carry on in case one or more tasks fail. A typical example of this could be that you want to check whether software is installed or not. Let's look at the following example to install Java 11 if, and only if, Java 8 is not installed. In the `roles/java/tasks/main.yaml` file, we are going to enter the following code:

```
- name: Verify if Java8 is installed
  command: rpm -q java-1.8.0-openjdk
  args:
    warn: False
  register: java
  ignore_errors: True
  changed_when: java is failed

- name: Ensure that Java11 is installed
  yum:
    name: java-11-openjdk
    state: present
  become: True
  when: java is failed
```

Before going forward with the other parts that are needed to execute this role, I'd like to spend a few words on the various parts of the role task list, since there are many new things.

In this task, we will execute an `rpm` command:

```
- name: Verify if Java8 is installed
  command: rpm -q java-1.8.0-openjdk
  args:
    warn: False
  register: java
  ignore_errors: True
  changed_when: java is failed
```

This code can have two possible output:

- Fail
- Return the complete name of the JDK package

Since we want only to check whether the package exists or not and then to go forward, we register the output (the *fifth* line) and ignore eventual failures (the *sixth* line).

When it fails, it means that `Java8` is not installed, and therefore we can proceed to install `Java11`:

```
- name: Ensure that Java11 is installed
  yum:
    name: java-11-openjdk
    state: present
  become: True
  when: java is failed
```

After we create the role, we will need the `hosts` file containing the host machine; in my case, it will be the following:

```
ws01
```

We will also need a playbook to apply the role, placed in `playbooks/hosts/j01.fale.io.yaml` and with the following content:

```
- hosts: ws01
  roles:
    - java
```

We can now execute it with the following:

ansible-playbook playbooks/hosts/ws01.yaml

We will get the following result:

```
PLAY [ws01] **********************************************************

TASK [Gathering Facts] **********************************************
ok: [ws01]

TASK [java : Verify if Java8 is installed] *************************
fatal: [ws01]: FAILED! => {"changed": true, "cmd": ["rpm", "-q",
"java-1.8.0-openjdk"], "delta": "0:00:00.028358", "end": "2019-02-10
10:56:22.474350", "msg": "non-zero return code", "rc": 1, "start":
"2019-02-10 10:56:22.445992", "stderr": "", "stderr_lines": [], "stdout":
"package java-1.8.0-openjdk is not installed", "stdout_lines": ["package
java-1.8.0-openjdk is not installed"]}
...ignoring

TASK [java : Ensure that Java11 is installed] **********************
changed: [ws01]

PLAY RECAP **********************************************************
ws01 : ok=3 changed=2 unreachable=0 failed=0
```

As you can see, the installation check failed since Java was not installed on the machine, and for this reason, the other task has been executed as expected.

Trigger failure

There are cases when you want to trigger a failure directly. This can happen for multiple reasons, even if there are disadvantages in doing so since, when you trigger the failure, the playbook will be brutally interrupted and this could leave your machine in an inconsistent state if you are not careful. One case where I have seen it work very well is when you are running a non-idempotent playbook (for instance, building a newer version of an application) and you need a variable (for instance, the version/branch to deploy) set. In this case, you can check that the expected variable is correctly configured before starting to run the operations to ensure that everything will work as expected later on.

Let's put the following code in `playbooks/maven_build.yaml`:

```
- hosts: all
  tasks:
    - name: Ensure the tag variable is properly set
      fail: 'The version needs to be defined. To do so, please add: --
extra-vars "version=$[TAG/BRANCH]"'
      when: version is not defined
    - name: Get last Project version
      git:
        repo: https://github.com/org/project.git
        dest: "/tmp"
        version: '{{ version }}'
    - name: Maven clean install
      shell: "cd /tmp/project && mvn clean install"
```

As you can see, we expect the user to add `--extra-vars "version=$[TAG/BRANCH]"` in the script-calling command. We could have put a branch to use by default, but this is too risky because the user may lose focus and forget to add the right branch name themselves, which would lead to compiling (and deploying) the wrong version of the application. The `fail` module also allows us to specify a message that will be displayed to the user.

> I think that the `fail` task is far more useful in playbooks that are run manually since, when a playbook is automatically run, managing the exception is often better than failing.

Using the `fail` module, you will be able to quit from playbooks as soon as a problem is detected.

Summary

In this chapter, we have seen how to debug Ansible playbooks using syntax checking, the checking mode with and without `--diff`, and functional testing.

As part of functional testing, we have seen how to perform assertions on the end state of the system, how to leverage tags for testing as well as how to use the `--syntax-check` option and the `ANSIBLE_KEEP_REMOTE_FILES` and `ANSIBLE_DEBUG` flags. Then, we moved to the management of failures, and, lastly, we saw how to trigger failures intentionally.

In the next chapter, we will discuss multi-tier environments as well as deployment methodologies.

Complex Environments

9

So far, we've seen how you can develop Ansible playbooks and test them. The final aspect is how to release playbooks into production. In most cases, you will have multiple environments to deal with before the playbook is released into production. This is similar to software that your developers have written. Many companies have multiple environments, and usually your playbook will follow these steps:

- Development environment
- Testing environment
- Staging environment
- Production

Some companies name those environments in different ways, and some companies have additional environments, such as the certification environment where all software has to be certified before it can go to production.

When you write your playbooks and set up roles, we strongly recommend that you keep in mind the notion of the environments right from the start. It might be worthwhile to talk to your software and operations teams to figure out exactly how many environments your setup has to cater to. We'll list a couple of approaches with examples that you can follow in your environment.

The following topics will be covered in this chapter:

- Code based on the Git branch
- Software distribution strategy
- Deploying a web app with a revision control system
- Deploying a web app with RPM packages
- Building compiled software with RPM packaging

Technical requirements

To be able to follow the examples in this chapter, you will need a UNIX machine capable of building RPM packages. My suggestion would be a Fedora or CentOS installation (either bare metal or in a virtual machine).

You can download all the files from this book's GitHub repository at `https://github.com/PacktPublishing/Learning-Ansible-2.X-Third-Edition/tree/master/Chapter09`.

Code based on the Git branch

Let's assume you have four environments to take care of, which are as follows:

- Development
- Testing
- Stage
- Production

In the Git branch-based method, you will have one environment per branch. You will always make changes to **Development** first, and then promote those changes to **Testing** (merge or cherry-pick, and tag commits in Git), **Stage**, and **Production**. In this approach, you will hold one single inventory file, one set of variable files, and, finally, a bunch of folders dedicated to roles and playbooks per branch.

A single stable branch with multiple folders

In this approach, you will always maintain the development and master branches. The initial code is committed to the development branch, and once stable, you will promote it to the master branch. The same roles and playbooks that exist in the master branch will run across all environments. On the other hand, you will have separate folders for each of your environments. Let's look at an example. We'll show how you can have a separate configuration and an inventory for two environments: staging and production. You can extend it to your scenario to fit all the environments you use. First, let's look at the playbook in `playbooks/variables.yaml` that will run across these multiple environments and has the following content. Full code is available on GitHub:

```
- hosts: web
  user: vagrant
  tasks:
    - name: Print environment name
```

```
      debug:
        var: env
    - name: Print db server url
      debug:
        var: db_url
    - name: Print domain url
      debug:
        var: domain
...
```

As you can see, there are two sets of tasks in this playbook:

- Tasks that run against DB servers
- Tasks that run against web servers

There is also an extra task to print the environment name that is common to all servers in a particular environment. We will also have two different inventory files.

The first one will be called `inventory/production` and will have the following content:

```
[web]
ws01.fale.io
ws02.fale.io

[db]
db01.fale.io

[production:children]
db
web
```

The second one will be called `inventory/staging` and will have the following content:

```
[web]
ws01.staging.fale.io
ws02.staging.fale.io

[db]
db01.staging.fale.io

[staging:children]
db
web
```

As you can see, we have two machines for the web section and one for the db in each environment. Furthermore, we have a different set of machines for stage and production environments. The additional section, [ENVIRONMENT:children], allows you to create a group of groups. This would mean that any variables that are defined in the ENVIRONMENT section will apply to both the db and web groups, unless they're overridden in the individual sections, respectively. The next interesting part would be to look at variable values for each of the environments and see how they are separated out in each environment.

Let's start with the variables that will be the same for all our environments, located in inventory/group_vars/all:

```
db_user: mysqluser
```

The only variable that is the same for both our environments is db_user.

We can now look at the production-specific variables, located in inventory/group_vars/production:

```
env: production
domain: fale.io
db_url: db.fale.io
db_pass: this_is_a_safe_password
```

If we now look at the stage-specific variables located in inventory/group_vars/staging, we will find the same variables we had in the production one, but with different values:

```
env: staging
domain: staging.fale.io
db_url: db.staging.fale.io
db_pass: this_is_an_unsafe_password
```

We can now validate that we received the expected results. First, we are going to run against the staging environment:

```
ansible-playbook -i inventory/staging playbooks/variables.yaml
```

We should receive an output similar to the following. Full code output is available in GitHub:

```
PLAY [web] ********************************************************

TASK [Gathering Facts] *******************************************
ok: [ws01.staging.fale.io]
ok: [ws02.staging.fale.io]

TASK [Print environment name] ************************************
ok: [ws01.staging.fale.io] => {
    "env": "staging"
}
ok: [ws02.staging.fale.io] => {
    "env": "staging"
}

TASK [Print db server url] ***************************************
ok: [ws01.staging.fale.io] => {
    "db_url": "db.staging.fale.io"
}
ok: [ws02.staging.fale.io] => {
    "db_url": "db.staging.fale.io"
}

...
```

We can now run against the production environment:

```
ansible-playbook -i inventory/production playbooks/variables.yaml
```

We will receive the following result:

```
PLAY [web] ********************************************************

TASK [Gathering Facts] *******************************************
ok: [ws02.fale.io]
ok: [ws01.fale.io]

TASK [Print environment name] ************************************
ok: [ws01.fale.io] => {
 "env": "production"
}
ok: [ws02.fale.io] => {
 "env": "production"
}

TASK [Print db server url] ***************************************
```

```
  ok: [ws01.fale.io] => {
   "db_url": "db.fale.io"
  }
  ok: [ws02.fale.io] => {
   "db_url": "db.fale.io"
  }

  . . .
```

You can see that the Ansible run picked up all of the relevant variables that were defined for the staging environment.

If you're using this approach to gain a stable master branch for multiple environments, it's best to use a mix of environment-specific directories, group_vars, and inventory groups to tackle the scenario.

Software distribution strategy

Deploying applications is probably one of the most complex tasks in the **Information and Communication Technology (ICT)** field. This is mainly caused by the fact that it often requires changing the state of the majority of machines that are somehow part of that application. In fact, often, you find yourself having to change the state of load balancers, distribution servers, application servers, and database servers all at the same time during a deployment. New technologies, such as containers, are trying to make those operations simpler, but often it is not easy or possible to just move a legacy application to a container.

We are now going to look at the various software distribution strategies and how Ansible can help with each one.

Copying files from the local machine

This is probably the oldest strategy to distribute software. The idea is to have the files on the local machine (often used to develop the code) and as soon as the change is made, a copy of the file is put on the server (usually via FTP). This way of deploying code was often used for web development, where the code (usually in PHP) does not need any compilation.

This distribution strategy should be avoided due to its multiple problems:

- It's very hard to rollback.
- It's impossible to track changes to the various deployments.
- There's no deployment history.
- It's easy to make errors during the deployment.

Although this distribution strategy can be very easily automated with Ansible, I strongly suggest that you immediately move to a different strategy that allows you to have a safer distribution strategy.

Revision control system with branches

Many companies are using this technique to distribute their software, mainly for uncompiled software. The idea behind this technique is to set up your server to use a local copy of your code repository. With SVN, this was possible but not very easy to manage properly, while Git allowed a simplification of this technique, making it very popular.

This technique has big advantages over the one we have just seen; the main ones are as follows:

- Easy rollbacks
- Very easy to obtain the history of changes
- Very easy deployments (mainly if Git is used)

On the other hand, this technique still has multiple disadvantages:

- No deployment history
- Hard for compiled software
- Possible security problems

I'd like to discuss the possible security problems you may encounter with this technique a little bit more. What can be very tempting is to download your Git repository directly in the folder that you use to distribute the content, so if it's a web server, this would be the `/var/www/` folder. This has obvious advantages, since to deploy you'll only need to perform a `git pull`. The disadvantage is that Git will create the `/var/www/.git` folder, which will contain your entire Git repository (history included) and, if not properly protected, will be freely downloadable by anyone.

About 1% of Alexa's top 1 million websites have the Git folder publicly accessible, so be very careful if you want to use this distribution strategy.

Revision control system with tags

Another way of using revision control systems that is a little bit more complex but that has some advantages is leveraging the tagging system. This method requires you to tag every time a new deployment has to be done and then check the specific tag on the server.

This has all the advantages of the previous method, with the addition of the deployment history. The compiled software problem and possible security problems are the same as in the previous method.

RPM packages

A very common way to deploy software (mainly for compiled applications, but also advantageous for non-compiled applications) is using some kind of packaging system. Some languages, such as Java, have included systems (WAR, in Java's case), but there are also packaging systems that can be used for any kind of applications, such as **RPM Package Manager** (**RPM**). RPM was originally developed by Erik Troan and Marc Ewing, and released in 1997 for Red Hat Linux. Since then, it has been adopted by many Linux distributions and is one of the two main ways to distribute software in the Linux world, with the other being DEB. The disadvantage of these systems is that they are a little bit more complex than the previous methods, but those systems can grant a higher level of security, as well as versioning. Also, these systems are easily injectable in a CI/CD pipeline, so the real complexity is much lower than what it could seem at first sight.

Preparing the environment

To see how we can deploy the code in the ways we talked about in the *Software distribution strategy* section, we will need an environment, and obviously we are going to create it using Ansible. First of all, to ensure that our roles are properly loaded, we need the ansible.cfg file with the following content:

```
[defaults]
roles_path = roles
```

Then, we need the `playbooks/groups/web.yaml` file to allow us to properly bootstrap our web servers:

```
- hosts: web
  user: vagrant
  roles:
    - common
    - webserver
```

As you can imagine from the previous file content, we will need to create the roles `common` and `webserver`, which are very similar to the ones we created in Chapter 4, *Handling Complex Deployment*. We will start with the `roles/common/tasks/main.yaml` file with the following content. The full code is available on GitHub:

```
- name: Ensure EPEL is enabled
  yum:
    name: epel-release
    state: present
  become: True
- name: Ensure libselinux-python is present
  yum:
    name: libselinux-python
    state: present
  become: True
- name: Ensure libsemanage-python is present
  yum:
    name: libsemanage-python
    state: present
  become: True
...
```

Here is its `motd` template in `roles/common/templates/motd`:

```
              This system is managed by Ansible
   Any change done on this system could be overwritten by Ansible

OS: {{ ansible_distribution }} {{ ansible_distribution_version }}
Hostname: {{ inventory_hostname }}
eth0 address: {{ ansible_eth0.ipv4.address }}

           All connections are monitored and recorded
    Disconnect IMMEDIATELY if you are not an authorized user
```

We can now move to the `webserver` role—more specifically, to the `roles/webserver/tasks/main.yaml` file. Full code file is available in GitHub:

```
---
- name: Ensure the HTTPd package is installed
  yum:
    name: httpd
    state: present
  become: True
- name: Ensure the HTTPd service is enabled and running
  service:
    name: httpd
    state: started
    enabled: True
  become: True
- name: Ensure HTTP can pass the firewall
  firewalld:
    service: http
    state: enabled
    permanent: True
    immediate: True
  become: True
...
```

We also need to create the handler in `roles/webserver/handlers/main.yaml` with this content:

```
---
- name: Restart HTTPd
  service:
    name: httpd
    state: restarted
  become: True
```

We add the following content to the `roles/webserver/templates/index.html.j2` file:

```
<html>
    <body>
        <h1>Hello World!</h1>
        <p>This page was created on {{ ansible_date_time.date }}.</p>
        <p>This machine can be reached on the following IP addresses</p>
        <ul>
{% for address in ansible_all_ipv4_addresses %}
            <li>{{ address }}</li>
{% endfor %}
        </ul>
    </body>
</html>
```

Lastly, we need to touch the `roles/webserver/files/website.conf` file, leaving it empty for now, but it needs to exist.

We can now provision a couple of CentOS machines (I provisioned `ws01.fale.io` and `ws02.fale.io`) and ensure that the inventory is right. We can configure those machines by running their group playbook:

```
ansible-playbook -i inventory/production playbooks/groups/web.yaml
```

We will receive the following output. Full code output is available on GitHub:

```
PLAY [web] **********************************************************

TASK [Gathering Facts] *********************************************
ok: [ws01.fale.io]
ok: [ws02.fale.io]

TASK [common : Ensure EPEL is enabled] ****************************
ok: [ws02.fale.io]
ok: [ws01.fale.io]

TASK [common : Ensure libselinux-python is present] *****************
ok: [ws02.fale.io]
ok: [ws01.fale.io]

TASK [common : Ensure libsemanage-python is present] ****************
ok: [ws01.fale.io]
ok: [ws02.fale.io]

TASK [common : Ensure we have last version of every package] *********
changed: [ws02.fale.io]
changed: [ws01.fale.io]

. . .
```

We can now point our browser to our nodes on port 80 to check that the HTTPd page is displayed as expected. Now that we have the basic webserver up and running, we can now focus on deploying a web application.

Deploying a web app with a revision control system

We are now going to perform our first deployment of the web application from a revision control system (Git) directly to our server using Ansible. So, we are going to deploy a simple PHP application that will be composed of only a single PHP page. The source is available at the following repository: `https://github.com/Fale/demo-php-app`.

To deploy it, we will need the following code placed in `playbooks/manual/rcs_deploy.yaml`:

```
- hosts: web
  user: vagrant
  tasks:
    - name: Ensure git is installed
      yum:
        name: git
        state: present
      become: True
    - name: Install or update website
      git:
        repo: https://github.com/Fale/demo-php-app.git
        dest: /var/www/application
      become: True
```

We can now run the deployer with the following command:

```
ansible-playbook -i inventory/production/playbooks/manual/rcs_deploy.yaml
```

This is the expected result:

```
PLAY [web] ********************************************************

TASK [Gathering Facts] *******************************************
ok: [ws02.fale.io]
ok: [ws01.fale.io]

TASK [Ensure git is installed] ***********************************
changed: [ws01.fale.io]
changed: [ws02.fale.io]

TASK [Install or update website] *********************************
changed: [ws02.fale.io]
changed: [ws01.fale.io]

PLAY RECAP *******************************************************
```

```
ws01.fale.io                    : ok=3 changed=2 unreachable=0 failed=0
ws02.fale.io                    : ok=3 changed=2 unreachable=0 failed=0
```

At the moment, our application is not yet reachable, since we have no HTTPd rule for that folder. To achieve this, we will need to change the `roles/webserver/files/website.conf` file with the following content:

```
<VirtualHost *:80>
    ServerName app.fale.io
    DocumentRoot /var/www/application
    <Directory /var/www/application>
        Options None
    </Directory>
    <DirectoryMatch ".git*">
        Require all denied
    </DirectoryMatch>
</VirtualHost>
```

As you can see, we are just displaying this application to the users who are reaching our server with the `app.fale.io` URL and not to everyone. This will ensure that all your users will have a consistent experience. Also, you can see that we are blocking all access to the `.git` folder (and all its content). This is needed for the security reasons we mentioned in the *Software distribution strategy* section of this chapter.

We can now re-run the web playbook to ensure that our HTTPd configuration gets propagated:

```
ansible-playbook -i inventory/production playbooks/groups/web.yaml
```

This is the result we are going to receive. Full code output is available on GitHub:

```
PLAY [web] ************************************************************

TASK [Gathering Facts] ***********************************************
ok: [ws01.fale.io]
ok: [ws02.fale.io]

TASK [common : Ensure EPEL is enabled] *****************************
ok: [ws02.fale.io]
ok: [ws01.fale.io]

TASK [common : Ensure libselinux-python is present] ***************
ok: [ws01.fale.io]
ok: [ws02.fale.io]

TASK [common : Ensure libsemanage-python is present] **************
ok: [ws01.fale.io]
```

```
ok: [ws02.fale.io]
```

`. . .`

You can now check and see that everything works properly.

We have seen how we can get a source from Git and deploy it to a web server so that it's promptly available to our users. We are now going to dive into another distribution strategy: deploying a web app with RPM packages.

Deploying a web app with RPM packages

To deploy an RPM package, we will need to create it in the first place. To do so, the first thing we need is a **SPEC file**.

Creating a SPEC file

The first thing we need to do is create a SPEC file, which is a recipe for instructing rpmbuild on how to actually create the RPM package. We are going to locate the SPEC file in spec/demo-php-app.spec. Following is the snippet content and the full code is available on GitHub:

```
%define debug_package %{nil}
%global commit0 b49f595e023e07a8345f47a3ad62a6f50f03121e
%global shortcommit0 %(c=%{commit0}; echo ${c:0:7})

Name: demo-php-app
Version: 0
Release: 1%{?dist}
Summary: Demo PHP application

License: PD
URL: https://github.com/Fale/demo-php-app
Source0: %{url}/archive/%{commit0}.tar.gz#/%{name}-%{shortcommit0}.tar.gz

%description
This is a demo PHP application in RPM format
  . . .
```

Let's see what the various parts do and mean before moving on:

```
%define debug_package %{nil}
%global commit0 b49f595e023e07a8345f47a3ad62a6f50f03121e
%global shortcommit0 %(c=%{commit0}; echo ${c:0:7})
```

Those first three lines are variables declarations.

The first one will disable the generation of a debug package. By default, `rpmbuild` will create a debug package every time and include all debugging symbols, but in this case we don't have any debugging symbols, since we are not making any compilation.

The second puts the hash of the commit in the `commit0` variable. The third one calculates the value of `shortcommit0`, which is calculated as the first eight characters of the `commit0` string:

```
Name:        demo-php-app
Version:     0
Release:     1%{?dist}
Summary:     Demo PHP application

License:     PD
URL:         https://github.com/Fale/demo-php-app
Source0:     %{url}/archive/%{commit0}.tar.gz#/%{name}-
%{shortcommit0}.tar.gz
```

In the first line, we declare the name, version, release number, and summary. The difference between version and release is that the version is the upstream version, while the release is the SPEC version for that upstream release.

The license is the source license, not the SPEC license. The URL is used to track the upstream website. The `source0` field is used by `rpmbuild` to find out how the source file is called (if more than one file is present, we can user `source1`, `source2`, and so on). Also, if the source fields are valid URI, we can use the `spectool` to download them automatically. This is the `description` of the software that's packaged in the RPM package:

```
%description
This is a demo PHP application in RPM format
```

The `prep` phase is the one where the source(s) get uncompressed and eventually patch(es) are applied. The `%autosetup` will `uncompress` the first source, as well as apply all patches. In this part, you can also perform other operations that need to be executed before the `build` phase and have the goal of preparing the environment for the `build` phase:

```
%prep
%autosetup -n %{name}-%{commit0}
```

Here, we would put all actions of the `build` phase. In our case, our sources do not need to be compiled, and therefore it is empty:

```
%build
```

In the `install` phase, we put the files in the `%{buildroot}` folder, which will mimic the target filesystem:

```
%install
mkdir -p %{buildroot}/var/www/application
ls -alh
cp index.php %{buildroot}/var/www/application
```

The `files` section is needed to declare which files are to be put in the package:

```
%files
%dir /var/www/application
/var/www/application/index.php
```

The `changelog` is needed to track who released a new version when and with which changes:

```
%changelog
* Sun Feb 24 2019 Fabio Alessandro Locati - 0.1
- Initial packaging
```

Now that we have the SPEC file, we need to build it. To do so, we could use a production machine, but this would increase the attack surface to that machine, so it's better to avoid it. There are multiple ways to build your RPM software. The four main ways are as follows:

- Manually
- Automate the manual way with Ansible
- Jenkins
- Koji

Let's look at the differences very briefly.

Building RPMs manually

The simplest way to build an RPM package is doing so in a manual way.

The big advantage is that you need only a few, simple steps to install `build`, and for this reason many people who are starting with RPM start from here. The disadvantage is that the process will be manual, and therefore human errors can spoil the result. Also, a manual build is very difficult to audit, since the only auditable part is the output and not the process itself.

To build RPM packages, you will need a Fedora or an EL (Red Hat Enterprise Linux, CentOS, Scientific Linux, Oracle Enterprise Linux) system. If you are using Fedora, you will need to execute the following command to install all the necessary software:

```
sudo dnf install -y fedora-packager
```

If you are running an EL system, the command you'll need to execute is as follows:

```
sudo yum install -y mock rpm-build spectool
```

In either case, you'll need to add the user you'll use to the `mock` group, and to do so, you need to execute the following:

```
sudo usermod -a -G mock [yourusername]
```

 Linux loads the users at login, so to apply a group change, you need to restart your session.

At this point, we can copy the SPEC file into the folder (usually, `$HOME` is a good one) and perform the following actions:

```
mkdir -p ~/rpmbuild/SOURCES
```

This will create the `$HOME/rpmbuild/SOURCES` folder that is needed in the process. The `-p` option will automatically create all folders in the path that are missing. We used `spectool` to download the source file and place it in the appropriate directory. The `spectool` will automatically get the URL from the SPEC file so that we don't have to remember it:

```
spectool -R -g demo-php-app.spec
```

We now need to create an `src.rpm` file, and to do so, we can use `rpmbuild`:

```
rpmbuild -bs demo-php-app.spec
```

This command will output something like this:

```
Wrote: /home/fale/rpmbuild/SRPMS/demo-php-app-0-1.fc28.src.rpm
```

Some small differences in the name could be present; for instance, you will probably have a different $HOME folder and you could have something other than fc24, if you are using something different than Fedora 24 to build the package. At this point, we can create the binary file with the following code:

```
mock -r epel-7-x86_64 /home/fale/rpmbuild/SRPMS/demo-php-
app-0-1.fc28.src.rpm
```

Mock allows us to build RPM packages in a clean environment and also, thanks to the -r option, it allows us to build for different versions of Fedora, EL, and Mageia. This command will give you a very long output, which we'll not cover here, but in the last few lines, there is useful information. If everything is built properly, this is the last few lines you should see:

```
Wrote: /builddir/build/RPMS/demo-php-app-0-1.el7.centos.x86_64.rpm
Executing(%clean): /bin/sh -e /var/tmp/rpm-tmp.d4vPhr
+ umask 022
+ cd /builddir/build/BUILD
+ cd demo-php-app-b49f595e023e07a8345f47a3ad62a6f50f03121e
+ /usr/bin/rm -rf /builddir/build/BUILDROOT/demo-php-
app-0-1.el7.centos.x86_64
+ exit 0
Finish: rpmbuild demo-php-app-0-1.fc28.src.rpm
Finish: build phase for demo-php-app-0-1.fc28.src.rpm
INFO: Done(/home/fale/rpmbuild/SRPMS/demo-php-app-0-1.fc28.src.rpm)
Config(epel-7-x86_64) 0 minutes 58 seconds
INFO: Results and/or logs in: /var/lib/mock/epel-7-x86_64/result
Finish: run
```

The second-to-last line contains the path where you can find the results. If you look in that folder, you should find the following files:

```
drwxrwsr-x. 2 fale mock 4.0K Feb 24 12:26 .
drwxrwsr-x. 4 root mock 4.0K Feb 24 12:25 ..
-rw-rw-r--. 1 fale mock 4.6K Feb 24 12:26 build.log
-rw-rw-r--. 1 fale mock 3.3K Feb 24 12:26 demo-php-
app-0-1.el7.centos.src.rpm
-rw-rw-r--. 1 fale mock 3.1K Feb 24 12:26 demo-php-
app-0-1.el7.centos.x86_64.rpm
-rw-rw-r--. 1 fale mock 184K Feb 24 12:26 root.log
-rw-rw-r--. 1 fale mock  792 Feb 24 12:26 state.log
```

The three log files are very useful in case of problems during the compilation. The `src.rpm` file will be a copy of the `src.rpm` file we created with the first command, while the `x86_64.rpm` file is the mock one we created and the one we will need to install on our machines.

Building RPMs with Ansible

Since doing all of those steps manually can be long, boring, and error-prone, we can automate them with Ansible. The resulting playbook will probably not be the cleanest one, but will be able to execute all operations in a repeatable way.

For this reason, we are going to build a new machine from scratch. I'll call this machine `builder01.fale.io`, and we are also going to change the `inventory/production` file to match this change:

```
[web]
ws01.fale.io
ws02.fale.io

[db]
db01.fale.io

[builders]
builder01.fale.io

[production:children]
db
web
builders
```

Before diving into the `builders` role, we will need to do a couple of changes to the `webserver` roles to enable a new repository. The first is adding a task in `roles/webserver/tasks/main.yaml` to the end of the file, with the following code:

```
- name: Install our private repository
  copy:
    src: privaterepo.repo
    dest: /etc/yum.repos.d/privaterepo.repo
  become: True
```

The second change is actually creating the `roles/webserver/files/privaterepo.repo` file with the following content:

```
[privaterepo]
name=Private repo that will keep our apps packages
baseurl=http://repo.fale.io/
skip_if_unavailable=True
gpgcheck=0
enabled=1
enabled_metadata=1
```

We can now execute the `webserver` group playbook to make the changes effective with the following command:

```
ansible-playbook -i inventory/production playbooks/groups/web.yaml
```

The following output should appear. Full code output is available on GitHub:

```
PLAY [web] *********************************************************

TASK [Gathering Facts] ********************************************
ok: [ws02.fale.io]
ok: [ws01.fale.io]

TASK [common : Ensure EPEL is enabled] ***************************
ok: [ws02.fale.io]
ok: [ws01.fale.io]

TASK [common : Ensure libselinux-python is present] *************
ok: [ws01.fale.io]
ok: [ws02.fale.io]

TASK [common : Ensure libsemanage-python is present] ***********
ok: [ws01.fale.io]
ok: [ws02.fale.io]


. . .
```

As expected, the only change has been the deployment of our newly generated repository file.

We also need to create a role for `builders` with a `tasks` file located in `roles/builder/tasks/main.yaml` with the following content. The full code is available on GitHub:

```
- name: Ensure needed packages are present
  yum:
    name: '{{ item }}'
    state: present
  become: True
  with_items:
    - mock
    - rpm-build
    - spectool
    - createrepo
    - httpd

- name: Ensure the user ansible is in the mock group
  user:
    name: ansible
    groups: mock
    append: True
  become: True

...
```

Also, as part of the `builder` role, we need the `roles/builder/handlers/main.yaml` handler file with the following content:

```
- name: Restart HTTPd
  service:
    name: httpd
    state: restarted
  become: True
```

As you can guess from the `tasks` file, we will also need the `roles/builder/files/repo.conf` file with the following content:

```
<VirtualHost *:80>
    ServerName repo.fale.io
    DocumentRoot /var/www/repo
    <Directory /var/www/repo>
        Options Indexes FollowSymLinks
    </Directory>
</VirtualHost>
```

We also need a new `group` playbook in `playbooks/groups/builders.yaml` with the following content:

```
- hosts: builders
  user: vagrant
  roles:
    - common
    - builder
```

We can now create the host itself with the following:

```
ansible-playbook -i inventory/production playbooks/groups/builders.yaml
```

We are expecting a result similar to the following:

```
PLAY [builders] ****************************************************

TASK [Gathering Facts] ********************************************
ok: [builder01.fale.io]

TASK [common : Ensure EPEL is enabled] ****************************
changed: [builder01.fale.io]

TASK [common : Ensure libselinux-python is present] ***************
ok: [builder01.fale.io]

TASK [common : Ensure libsemanage-python is present] **************
changed: [builder01.fale.io]

TASK [common : Ensure we have last version of every package] ******
changed: [builder01.fale.io]

TASK [common : Ensure NTP is installed] ***************************
changed: [builder01.fale.io]

TASK [common : Ensure the timezone is set to UTC] *****************
changed: [builder01.fale.io]

...
```

Now that we have all the parts of the infrastructure ready, we can create the `playbooks/manual/rpm_deploy.yaml` file with the following content. The full code is available on GitHub:

```
- hosts: builders
  user: vagrant
  tasks:
    - name: Copy SPEC file to user folder
```

```
      copy:
        src: ../../spec/demo-php-app.spec
        dest: /home/vagrant
    - name: Ensure rpmbuild exists
      file:
        name: ~/rpmbuild
        state: directory
    - name: Ensure rpmbuild/SOURCES exists
      file:
        name: ~/rpmbuild/SOURCES
        state: directory
  ...
```

As we discussed, this playbook has a lot of commands and shells that are not very clean. In the future, it may be possible to write a playbook with the same features but with modules. Most actions are the same, as we discussed in the previous section. The new actions are toward the end; in fact, in this case, we copy the generated RPM file to a specific folder, we invoke `createrepo` to generate a repository in that folder, and then we force all web servers to update the generated package to the last version.

 To ensure the security of your application, it is important that the repository is only accessible internally and not publicly.

We can now run the playbook with the following command:

```
ansible-playbook -i inventory/production playbooks/manual/rpm_deploy.yaml
```

We expect a result such as the following. Full code output is available on GitHub:

```
PLAY [builders] ****************************************************

TASK [setup] ******************************************************
ok: [builder01.fale.io]

TASK [Copy SPEC file to user folder] ******************************
changed: [builder01.fale.io]

TASK [Ensure rpmbuild exists] *************************************
changed: [builder01.fale.io]

TASK [Ensure rpmbuild/SOURCES exists] *****************************
changed: [builder01.fale.io]

TASK [Download the sources] ***************************************
changed: [builder01.fale.io]
```

```
TASK [Ensure no SRPM files are present] ******************************
changed: [builder01.fale.io]

TASK [Build the SRPM file] ***********************************************
changed: [builder01.fale.io]
. . .
```

Building RPMs with CI/CD pipelines

Although this is not covered in this book, in more complex cases, you may want to use a CI/CD pipeline to create and manage RPM packages. The two main pipelines are based on two different types of software:

- Koji
- Jenkins

The Koji software has been developed by the Fedora community and Red Hat. It is released under the terms of the LGPL 2.1 license. This is the pipeline that currently gets used by Fedora, CentOS, as well as many other companies and communities to create all their RPMs (both for official testing, also known as **scratch builds**). Koji, by default, is not triggered by commit; it needs to be called **manually** from a user (through a web interface or CLI). Koji will automatically download the last version of the SPEC file in Git, download the source from a side-cache (this is optional, but suggested) or from the original location, and trigger the mock build. Koji supports only a mock build due to the fact that it is the only system that allows consistent and repeatable builds. Koji can store all the output artifacts forever or for a limited amount of time, based on the configuration. This is to ensure a very high level of auditability.

Jenkins is one of the most-used CI/CD managers and can also be used for RPM pipelines. The big disadvantage is that it needs to be configured from scratch with the consequence that more time is required, but this means it has more flexibility. Also, a big advantage of Jenkins is that many companies already have an instance of Jenkins, and this makes it easier to set up and maintain the infrastructure, since you can reuse an installation you already have, and therefore you don't have to manage fewer systems overall.

Building compiled software with RPM packaging

RPM packaging is very useful for non-binary applications and close to a necessity for binary applications. This is also true because the difference in complexity is pretty low between a non-binary and a binary case. In fact, the build and the installation will work in exactly the same way. The only thing that will change is the SPEC file.

Let's look at the SPEC file that's needed to compile and package a simple `Hello World!` application written in C:

```
%global commit0 7c288b9d80a6ef525c0cca8a744b32e018eaa386
%global shortcommit0 %(c=%{commit0}; echo ${c:0:7})

Name:           hello-world
Version:        1.0
Release:        1%{?dist}
Summary:        Hello World example implemented in C

License:        GPLv3+
URL:            https://github.com/Fale/hello-world
Source0:        %{url}/archive/%{commit0}.tar.gz#/%{name}-
%{shortcommit0}.tar.gz

BuildRequires:  gcc
BuildRequires:  make

%description
The description for our Hello World Example implemented in C

%prep
%autosetup -n %{name}-%{commit0}

%build
make %{?_smp_mflags}

%install
%make_install

%files
%license LICENSE
%{_bindir}/hello

%changelog
* Sun Feb 24 2019 Fabio Alessandro Locati - 1.0-1
- Initial packaging
```

As you can see, it's very similar to the one we saw for the PHP demo application. Let's look at the differences.

Let's dive a little bit into the various parts of the SPEC file:

```
%global commit0 7c288b9d80a6ef525c0cca8a744b32e018eaa386
%global shortcommit0 %(c=%{commit0}; echo ${c:0:7})
```

As you can see, we don't have the line to disable the debug package. Every time you package a compiled application, you should let rpm create the debug symbols package so that in the case of crashes, it will be easier to debug and understand the problem.

The following part of the SPEC file is shown here:

```
Name:           hello-world
Version:        1.0
Release:        1%{?dist}
Summary:        Hello World example implemented in C

License:        GPLv3+
URL:            https://github.com/Fale/hello-world
Source0:        %{url}/archive/%{commit0}.tar.gz#/%{name}-
%{shortcommit0}.tar.gz
```

As you can see, the changes in this section are only due to the fact that the new package has a different name and URL, but they are not linked by the fact that this is a compilable application:

```
BuildRequires:  gcc
BuildRequires:  make
```

In the non-compiled application, we did not need any packages present at build time, while in this case we will need the make and the gcc (compiler) applications. Different applications could require different tools and/or libraries to be present on the system at build time:

```
%description
The description for our Hello World Example implemented in C

%prep
%autosetup -n %{name}-%{commit0}

%build
make %{?_smp_mflags}
```

The `description` is package-specific and is not influenced by the compilation of the package. In the same way, the `%prep` phase works.

In the `%build` phase, we now have to make `%{?_smp_mflags}`. This is needed to tell `rpmbuild` to actually run `make` to build our application. The `_smp_mflags` variable will include a set of parameters to optimize the compilation to be multi-thread:

```
%install
%make_install
```

During the `%install` phase, we will issue the `%make_install` command. This macro will call `make install` with a set of additional parameters to ensure that the libraries are located in the right folder, as well as the binaries and so forth:

```
%files
%license LICENSE
%{_bindir}/hello
```

In this case, we only need to place the `hello` binary that was located in the right folder of the `buildroot` during the `%install` phase, and also add the `LICENSE` file containing the license:

```
%changelog
* Sun Feb 24 2019 Fabio Alessandro Locati - 1.0-1
- Initial packaging
```

The `%changelog` is very similar to the other SPEC file we saw, since it is not influenced by the involvement of a compilation.

After you have completed this, you can place it in `spec/hello-world.spec` and tweak `playbooks/manual/rpm_deploy.yaml` by saving it into `playbooks/manual/hello_deploy.yaml` with the following code snippet. The full code is available on GitHub:

```
- hosts: builders
  user: vagrant
  tasks:
    - name: Copy SPEC file to user folder
      copy:
        src: ../../spec/hello-world.spec
        dest: /home/ansible
    - name: Ensure rpmbuild exists
      file:
        name: ~/rpmbuild
        state: directory
    - name: Ensure rpmbuild/SOURCES exists
```

```
        file:
          name: ~/rpmbuild/SOURCES
          state: directory
    ...
```

As you can see, the only thing that we changes is that all references to `demo-php-app` got replaced with `hello-world`. Run it with the following command:

```
ansible-playbook -i inventory/production playbooks/manual/hello_deploy.yaml
```

We are going to get the following result. Full code output is available on GitHub:

```
PLAY [builders] ************************************************

TASK [setup] **************************************************
ok: [builder01.fale.io]

TASK [Copy SPEC file to user folder] **************************
changed: [builder01.fale.io]

TASK [Ensure rpmbuild exists] *********************************
ok: [builder01.fale.io]

TASK [Ensure rpmbuild/SOURCES exists] *************************
ok: [builder01.fale.io]

TASK [Download the sources] ***********************************
changed: [builder01.fale.io]

TASK [Ensure no SRPM files are present] ***********************
changed: [builder01.fale.io]

TASK [Build the SRPM file] ************************************
changed: [builder01.fale.io]

TASK [Execute mock] *******************************************
changed: [builder01.fale.io]

    ...
```

> You could eventually create a playbook that accepts the name of the package to build as a parameter, so that you don't need a different playbook for every package.

Deployment strategies

We have seen how to distribute software in your environment,so now, we will speak about deployment strategies, that is, how to upgrade your application without your service suffering from it.

There are three different problems you might incur during an update:

- Downtime during the update rollout.
- The new version has problems.
- The new version seems to work, until it fails.

The first problem is known to every system administrator. During the update, you are probably going to restart some services, and for the time between the start and the end of the service, your application will not be available on that machine. To address this problem, you would need multiple machines with a smart load balancer in front that will remove specify nodes from the pool of available nodes just before the upgrade on that specific node is performed, to add them back later as soon as the node has been upgraded.

The second problem can be prevented in multiple ways. The cleanest one would be testing in the CI/CD pipeline. In fact, those kinds of problems are pretty easy to find with simple tests. This can also be prevented with the methods we are going to see soon.

The third problem is by far the most complex. Manytimes, even ones on a global scale, have been generated by these kinds of problems. Usually, the problem is that the new version has some performance problems or memory leaks. Since the majority of deployments are done in the period of least load of the servers, as soon as the load increases, a performance problem or memory leak could kill your servers.

 To be able to use those methods in a proper way, you have to be able to ensure that your software can accept rollbacks. There are cases where this is not possible (that is, a database table gets removed in an update). We will not discuss how to avoid this, since this is part of the development strategy, and is not related to Ansible.

To work around those problems, two common deployment patterns are used: the **canary deployment** and the **blue/green deployment**.

Canary deployment

The canary deployment is a technique that involves updating a small percentage of your machines (often 5%) to the new version and instructing the load balancers to send only an equivalent amount of traffic to it. This has several advantages:

- During the update, you never have less than 95% of the capacity
- If the new version completely fails, you lose 5% of the capacity
- Since the load balancer divides the traffic between your new and your old version, if the new version has problems, only 5% of your users will see the problem
- You only need to have 5% capacity more than your expected load

Canary deployment is able to prevent all three problems we mentioned with a very small overhead (5%) and with low cost in the case of rollback (5%). For those reasons, this technique is used a lot by huge companies. Often, to ensure a similar user experience to users that live close to one another, geography is used to choose whether the user is going to hit the old or the new version.

When the test seems to be a success, the percentage can be increased progressively until 100% is reached.

It's possible to implement a canary deployment in multiple ways in Ansible. The way I suggest is the cleanest one, that is, using the inventory files, so that you have something such as the following:

```
[web-main]
ws[00:94].fale.io

[web-canary]
ws[95:99].fale.io

[web:children]
web-main
web-canary
```

In this way, you can set all the variables on the web group (the variables are going to be the same, no matter what the version OS, or at least they should be), but you can run a playbook easily against the canary group, the main group, or both groups at the same time. Another option would be to create two different inventory files, one for the canary group and the other for the main group with the groups, having the same name so that variables are shared.

Blue/green deployment

Blue/green deployment is very different compared to canary deployment, and it has some advantages and some disadvantages. The main advantages are as follows:

- Easier to implement
- Allows quicker iterations
- All users get moved at the same time
- Rollbacks have no performance degradation

Among the disadvantages, the main ones are the fact that you need to have double the machines available than what your application requires. This disadvantage can be easily mitigated if the application is running on a cloud (either private, public, or hybrid) scaling up the application resources for the deployment and then scale them back down.

Implementing blue/green deployment in Ansible is very easy. The simplest way is to create two different inventories (one for blue and one for green) and then simply manage your infrastructure as if they are different environments, such as production, staging, development, and so on.

Optimizations

Sometimes, Ansible feels slow, mainly if you have a very long list of tasks to execute and/or if you have a huge amount of machines. There are multiple reasons for this, and ways to avoid it, and we are going to look at three of those ways.

Pipelining

One of the reasons Ansible is slow by default is that for every module execution and for every host, Ansible will perform the following actions:

- SSH handshake
- Execute the task
- Close the SSH connection

As you can see, this means that if you have 10 tasks to be executed on a single remote server, Ansible will open (and close) the connection 10 times. Since the SSH protocol is an encrypted protocol, this makes the SSH handshake an even longer process, since the two parts have to negotiate the cyphers every single time.

Ansible allows us to reduce the execution time drastically by initiating the connections at the beginning of the playbook and keeping them alive for the whole execution so that it does not need to re-open the connection at every task. Over the course of Ansible's life, this feature has changed names multiple times, as well as the way it's enabled. From version 1.5, it's been called **pipelining**, and the way to enable it is by adding the following line to your `ansible.cfg` file:

```
pipelining=True
```

The reason this feature is not enabled by default is that many distributions ship with the `requiretty` option in `sudo`. The pipelining mode in Ansible and the `requiretty` option in `sudo` conflict and will make your playbooks fail.

 If you want to enable the pipelining mode, ensure that the `sudo` `requiretty` mode is disabled on your target machines.

Optimizing with_items

If you want to execute similar operations multiple times, it's possible to repeat the same task multiple times with different parameters or use the `with_items` option. Aside from the fact that `with_items` makes your code easier to read and to follow, it could also improve your performance. An example is with the installation of packages (that is, the `apt`, `dnf`, `yum`, `package` modules) where Ansible will perform a single command if you use `with_items`, as opposed to a single command for each package if you don't. As you can imagine, this can help boost your performance.

Understanding what happens when your tasks are executed

Even after you have implemented the methods we just talked about to speed up the playbook's execution, you may still find that some tasks take a very long time. This is very common with some tasks, even if it's possible with many other modules. The modules that usually give you this problem are as follows:

- Packaging management (that is, `apt`, `dnf`, `yum`, `package`)
- Cloud machine creation (that is, `digital_ocean`, `ec2`)

The reason for this slowness is often non-Ansible specific. An example case could be if you used a packaging management module to update your machines. This requires downloading tens or hundreds of megabytes on every machine and installing a high quantity of software. A way to speed up this kind of operation is to have a local repository in your data center and have all your machines pointing to it instead of your distribution repositories. This will allow your machines to download at higher speed and without using the public connection that is often limited in bandwidth, or metered.

 It's often important to understand what the modules do in the background to optimize the playbook's execution.

In the cloud machine creation case, Ansible just performs an API call to the chosen cloud provider and waits for the machine to be ready. DigitalOcean machines can take up to one minute to be created (and other clouds much longer), so Ansible will wait for that amount of time. Some modules have an asynchronous mode to avoid this wait period, but you'll have to ensure that the machine is ready before using it; otherwise, the modules that use the created machine will fail.

Summary

In this chapter, we have seen how you can deploy an application with Ansible, as well as the various distribution and deployment strategies you can use. We also saw how to create RPM packages with Ansible and how to optimize the performance of Ansible using different methods.

In the next chapter, we will be looking at how to use Ansible on Windows machines, where to find roles written by other people and how to use them, and a user interface for Ansible.

Section 4: Deploying an Application with Ansible

4

This section explains how to manage Windows nodes from Ansible and how to leverage Ansible Galaxy and Ansible Tower to maximize your productivity.

This section contains the following chapters:

- Chapter 10, *Introducing Ansible for Enterprises*
- Chapter 11, *Getting Started with AWX*
- Chapter 12, *Working with AWX Users, Permissions, and Organizations*

10
Introducing Ansible for Enterprises

In the previous chapter, we saw how Ansible works and how to leverage it. We have gone through this whole book—up until now—on the assumptions that we were targeting Unix machines, that we were going to write all of our playbooks ourselves, and that the Ansible CLI was what were looking for. We will now move away from those assumptions to see how we can go beyond typical Ansible usage.

In this chapter, we'll explore the following topics:

- Ansible on Windows
- Ansible Galaxy
- Ansible Tower

Technical requirements

Aside from Ansible itself, to be able to follow the examples in this chapter on your machine, you will need a Windows box.

Ansible on Windows

Ansible version 1.7 started being able to manage Windows machines with a few basic modules. After the acquisition of Ansible by Red Hat, a lot of effort was put into this task by Microsoft and many other companies and people. By the time of the 2.1 release, Ansible's ability to manage Windows machines was close to being complete. Some modules have been extended to work seamlessly on Unix and Windows, while in other cases, the Windows logic was so different from Unix that new modules needed to be created.

 At the time of writing, using Windows as a control machine is not supported, though some users have tweaked the code and their environment to make it work.

The connection from the control machine to Windows machines is not made over SSH; instead, it's made over **Windows Remote Management** (**WinRM**). You can visit Microsoft's website for a detailed explanation and implementation:
`http://msdn.microsoft.com/en-us/library/aa384426(v=vs.85).aspx`.

On the control machine, once you've installed Ansible, it's important that you install WinRM. You can do so via `pip` with the following command:

```
pip install "pywinrm>=0.3.0"
```

 You may need to use `sudo` or the `root` account to execute this command.

On each of the remote Windows machines, you need to install PowerShell version 3.0 or higher. Ansible provides a couple of helpful scripts to set it up:

- WinRM (`https://github.com/ansible/ansible/blob/devel/examples/scripts/ConfigureRemotingForAnsible.ps1`)
- PowerShell 3.0 upgrade (`https://github.com/cchurch/ansible/blob/devel/examples/scripts/upgrade_to_ps3.ps1`)

You will also need to allow port `5986` via the firewall, as this is the default WinRM connection port, and make sure it is accessible from the command center.

To make sure you can access the service remotely, run a `curl` command:

```
curl -vk -d `` -u "$USER:$PASSWORD" "https://<IP>:5986/wsman".
```

If basic authentication works, you can start running commands. Once the setup is done, you're ready to start running Ansible! Let's run the equivalent of the Windows version of the `Hello, world!` program in Ansible by running `win_ping`. In order to do this, let's set up our credentials file.

This can be done using `ansible-vault`, as follows:

```
$ ansible-vault create group_vars/windows.yml
```

As we've already seen, `ansible-vault` will ask you to set `password`:

```
Vault password:
Confirm Vault password:
```

At this point, we can add the variables we need:

```
ansible_ssh_user: Administrator
ansible_ssh_pass: <password>
ansible_ssh_port: 5986
ansible_connection: winrm
```

Let's set up our `inventory` file, as follows:

```
[windows]
174.129.181.242
```

Followed by this, let's run `win_ping`:

```
ansible windows -i inventory -m win_ping --ask-vault-pass
```

Ansible will ask us for `Vault password` and then print the result of the run, as follows:

```
Vault password:
174.129.181.242 | success >> {
    "changed": false,
    "ping": "pong"
}
```

We have now seen how we can connect to a remote machine. Now, you can manage Windows machines in the same way that you can manage Unix machines. The only thing to be aware of is that, due to the huge differences between the Windows OS and Unix systems, not every Ansible module will work properly. For this reason, many Unix modules have been rewritten from scratch to have similar behaviors to the Unix modules, but with completely different implementations. A list of those modules can be found at `https://docs.ansible.com/ansible/latest/modules/list_of_windows_modules.html`.

Ansible Galaxy

Ansible Galaxy is a free website where you can download Ansible roles that have been developed by the community and kick-start your automation within minutes. You can share or review community roles so that others can easily find the most trusted roles on Ansible Galaxy. You can start using Ansible Galaxy by simply signing up with social media applications such as Twitter, Google, and GitHub or by creating a new account on the Ansible Galaxy website at `https://galaxy.ansible.com/` and downloading the required roles using the `ansible-galaxy` command, which ships with Ansible version 1.4.2 and higher.

> In case you want to host your own local Ansible Galaxy instance, you can do so by fetching the code from `https://github.com/ansible/galaxy`.

To download an Ansible role from Ansible Galaxy, use the following command:

```
ansible-galaxy install username.rolename
```

You can also specify a version, as follows:

```
ansible-galaxy install username.rolename[,version]
```

If you don't specify a version, then the `ansible-galaxy` command will download the latest available version. You can install multiple roles in two ways; firstly, by passing multiple role names separated by a space, as follows:

```
ansible-galaxy install username.rolename[,version]
username.rolename[,version]
```

Secondly, you can do so by specifying role names in a file and passing that filename to the `-r/--role-file` option. For instance, you could create the `requirements.txt` file with the following content:

```
user1.rolename,v1.0.0
user2.rolename,v1.1.0
user3.rolename,v1.2.1
```

You could then install roles by passing the filename to the `ansible-galaxy` command, as follows:

```
ansible-galaxy install -r requirements.txt
```

Let's see how we can use `ansible-galaxy` to download a role for Apache HTTPd:

```
ansible-galaxy install geerlingguy.apache
```

You'll see output similar to the following:

```
- downloading role 'apache', owned by geerlingguy
- downloading role from
https://github.com/geerlingguy/ansible-role-apache/archive/3.0.3.tar.gz
- extracting geerlingguy.apache to
/home/fale/.ansible/roles/geerlingguy.apache
- geerlingguy.apache (3.0.3) was installed successfully
```

The preceding `ansible-galaxy` command will download the Apache HTTPd role to the `~/.ansible/roles` directory. You can now directly use the preceding role in your playbook and create the `playbooks/galaxy.yaml` file with the following content:

```
- hosts: web
  user: vagrant
  become: True
  roles:
    - geerlingguy.apache
```

As you can see, we created a simple playbook with a `geerlingguy.apache` role. We can now test it:

```
ansible-playbook -i inventory playbooks/galaxy.yaml
```

This should give us the following output:

```
PLAY [web] ************************************************************

TASK [Gathering Facts] ***********************************************
ok: [ws01.fale.io]
```

```
TASK [geerlingguy.apache : Include OS-specific variables.] **********
ok: [ws01.fale.io]

TASK [geerlingguy.apache : Include variables for Amazon Linux.] ******
skipping: [ws01.fale.io]

TASK [geerlingguy.apache : Define apache_packages.] *****************
ok: [ws01.fale.io]

TASK [geerlingguy.apache : include_tasks] ***************************
included: /home/fale/.ansible/roles/geerlingguy.apache/tasks/setup-
RedHat.yml for ws01.fale.io

TASK [geerlingguy.apache : Ensure Apache is installed on RHEL.] ******
changed: [ws01.fale.io]

TASK [geerlingguy.apache : Get installed version of Apache.] ********
ok: [ws01.fale.io]

. . .
```

 As you may have noticed, many steps were skipped due to the fact that this role is designed to work on many different Linux distributions.

Now that you know how to leverage Ansible Galaxy roles, you can spend less time rewriting code that someone already wrote and spend more time writing the parts that are specific to your architecture and give you more value.

Pushing a role to Ansible Galaxy

Since Ansible Galaxy is a community-driven effort, you can also add your own roles to it. Before we can start the process of publishing it, we will need to prepare it.

Ansible gives us a tool to bootstrap a new Galaxy Role from a template. To leverage it, we can run the following command:

```
ansible-galaxy init ansible-role-test
```

This will create the `ansible-role-test` folder, along with all of the usual folders an Ansible Role usually has.

The only file that is going to be new to you is `meta/main.yaml`, which, even though it is possible to use without Ansible Galaxy, contains a lot of information about the role that's readable by Ansible Galaxy.

The main information that's available in this file that you will probably need to set accordingly to your needs are as follows:

- `author`: Your name.
- `description`: Put a description of the role here.
- `company`: Put the name of the company you work for here (or delete the line).
- `license`: Set the license that your module will have. Some suggested licenses are BSD (which is also the default), MIT, GPLv2, GPLv3, Apache, and CC-BY.
- `min_ansible_version`: Set the minimum version of Ansible that you've tested the role with.
- `galaxy_tags`: In this section, put the platforms and versions your module is written for.
- `dependencies`: List the roles that are required to execute your role.

To proceed with the publication, you need to log in to Galaxy using a GitHub account, and then you can go to **My Content** to start adding a content.

After you press **Add Content**, a window will appear that will show you the repositories that you can choose from, as shown in the following screenshot:

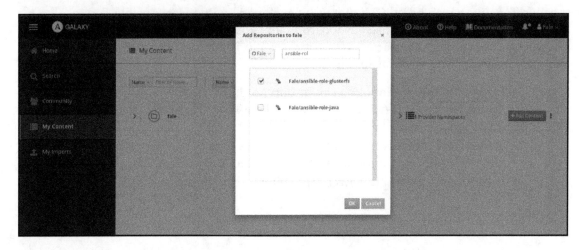

After selecting the right repository and then clicking the **OK** button, Ansible Galaxy will start to import the given role.

If you go back to the **My Content** page a few minutes after doing this, you'll see your roles and their status, like so:

You can now use the role in the same way as other people would have done. Remember to update it when changes are needed!

Ansible Tower and AWX

Ansible Tower is a web-based GUI that was developed by Red Hat. Ansible Tower provides you with an easy-to-use dashboard where you can manage your nodes and role-based authentication to control access to your Ansible Tower dashboard. The biggest features of Ansible Tower are as follows:

- **LDAP/AD integration**: You can import (and give privileges to) users based on the result of LDAP/AD queries that Ansible Tower performs on your LDAP/AD server.
- **Role-based access control**: It limits the users to only run the playbooks they are authorized to run and/or target only a limited amount of hosts.
- **REST API**: All Ansible Tower capabilities are exposed via a REST API.

- **Job scheduling**: Ansible Tower allows us to schedule jobs (playbook execution).
- **Graphical inventory management**: Ansible Tower manages the inventory in a more dynamic way than Ansible.
- **Dashboard**: Ansible Tower allows us to see the situation of all current and previous job executions.
- **Logging**: Ansible Tower logs all the results of every job execution to be able to go back and check if needed.

During the acquisition of Ansible Inc. by Red Hat, it was promised that Ansible Tower would have become open source. In 2017, this happened, and it came back with the name AWX.

AWX and Ansible Tower are commonly used in the Enterprise version due to the very handy features it provides to the Ansible ecosystem. We are going to discuss these in more detail in the chapters that follow.

Summary

In this chapter, we have seen how to move Ansible outside the Unix world by looking at how to control Windows hosts. We then moved to Ansible Galaxy, where you can find many roles written by other people that you can simply reuse. Lastly, we touched on Ansible Tower, which is an open source incarnation of AWX. In the upcoming chapters, we are going to discuss more about AWX, from the installation process to running your first job.

11
Getting Started with AWX

As we have seen in the previous chapters, Ansible is a very powerful tool. This is not enough to make it ubiquitous. In fact, for it to become ubiquitous, a tool needs to be easy to use at any user level and easy to integrate in various ways with existing environments.

Ansible Inc recognized this and created a tool called Ansible Tower, which was basically a Web UI and API set around Ansible. Ansible Tower was a closed source tool, which was also the main source of revenue for the company. When Red Hat announced that it had acquired Ansible, its management also committed to making Ansible Tower open source. A couple of years later, Red Hat open sourced Ansible Tower, creating the AWX project, which is now the upstream of Ansible Tower, in the same way Fedora is the upstream of Red Hat Enterprise Linux.

Before AWX, other Web UIs and API sets were developed in the open source community, such as Semaphore. AWX and Ansible Tower are not the only Web UI and API sets for Ansible today, but they are the more actively developed solutions.

In this chapter, we are going to see how to set up AWX and learn how to use it. More specifically, we are going to discuss the following:

- Setting up AWX
- Understanding what an AWX project is and how to leverage it
- Understanding what an AWX inventory is and how it differs from an Ansible inventory
- Understanding what an AWX job template is and how to create one
- Understanding what an AWX job is and how to execute your first job

Technical requirements

For this chapter, you will need a machine that can run `ansible` and `docker` and has `docker-py` installed.

Setting up AWX

Unlike Ansible, installing AWX involves more than a single command, but it's still fairly quick and easy.

First of all, you need to install `ansible`, `docker`, and `docker-py`. After this, you need to give permission to the desired user to run Docker. Lastly, you need to download AWX Git repo and execute an `ansible` playbook.

Installing Ansible, Docker, and Docker-py in Fedora

Let's begin by installing `docker`, `ansible`, and `docker-py` packages in Fedora:

```
sudo dnf install ansible docker python-docker-py
```

To start and enable the Docker service use the following:

```
sudo systemctl start docker
sudo systemctl enable docker
```

Now that we've installed `ansible`, `docker`, and `docker-py`, let's move on to granting user access to Docker.

Giving the current user permission to use Docker in Fedora

To ensure that the current user can use Docker (by default, Fedora only allows root to use it), you need to create a new Docker group, assign the current user to it, and restart Docker:

```
sudo groupadd docker && sudo gpasswd -a ${USER} docker && sudo systemctl
restart docker
```

Since the groups are assigned only at the beginning of the session, you would need to restart your session, but we can force Linux to add your new group to your current session, by executing the following:

```
newgrp docker
```

Now that we have all the prerequisites ready, we can move to the real AWX installation.

Installing AWX

The first thing that we need is to check out the `git` repository by executing the following:

```
git clone https://github.com/ansible/awx.git
```

As soon as Git has completed its task, we can change the directory to the one containing the installer and run it:

```
cd awx/installer/
ansible-playbook -i inventory install.yml
```

This will install AWX in Docker containers and default configurations. You can tweak the configurations (before running the last command) by changing the `inventory` file, in the same folder.

When the installation process is completed, you can open your browser and point to `https://localhost` and log in using the `admin` username with the `password` password.

After the login, you should see a page that resemble the following one:

Having set up AWX, you will now be able to execute Ansible playbooks without using the Ansible CLI anymore. To start this, we will firstly need a project, so let's see how to set it up.

Creating new AWX projects

AWX assumes that you save your playbooks somewhere, and to be able to use them in AWX, we need to create a project.

A project is basically the AWX placeholder for a repository containing Ansible resources (roles and playbooks).

When you go in the **Projects** section, in the left-hand menu bar, you'll see something like the following:

As you can see, a **Demo Project** is already in place (the installer created it for us!) and it is backed by a Git repository.

On the left-hand side of the project name, a white circle is present, and represents that the specific project has never been pulled. If a green circle were present, it would mean that the project has been pulled successfully. A pulsing green circle means that a pull is in progress, while a red stop sign means that something went wrong.

In the same line of the project, there are three buttons:

- **Get SCM last revision**: To fetch the current latest revision of the code
- **Duplicate**: To create a duplicate of the project
- **Delete**: To delete the project

In the top right-hand part of the card, you can notice a green plus button. This is the button that allows us to add more projects.

By selecting it, a new **NEW PROJECT** card appears on top of the **Projects** card, where you can add the new project.

The **NEW PROJECT** card will look like the following:

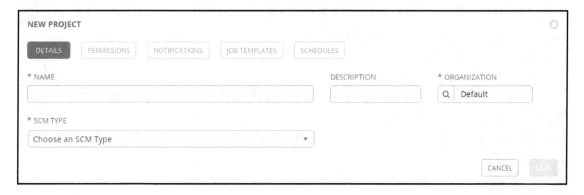

It's asking for information about the project you are going to create:

- **NAME**: This is the display name of your project. This is for human-usage, so make it human-sensible!
- **DESCRIPTION**: An additional display (still for humans) to make sense of the project goals.
- **ORGANIZATION**: The organization that is going to own the project. This will be covered in the next chapter. For now, let's leave the default.
- **SCM TYPE**: The type of SCM that your code is contained in. At the moment of writing, the supported options are: **Manual**, **Git**, **Mercurial**, **Subversion**, and **Red Hat Insights**.
- Based on the SCM type you chose, more fields will appear, such as the **SCM URL**, and the **SCM branch**.

When you complete all the mandatory fields, you can save and see that a new project has been added.

Using AWX inventories

AWX inventories are the equivalent of Ansible inventories in the AWX world. Since AWX is a graphical tool, inventories are not stored as files (as done in Ansible) but will be manageable with the AWX user interface. Not being tied to a file also gives more flexibility to AWX inventories compared to Ansible inventories.

AWX has different ways to manage the inventories.

You can reach this by clicking the **Inventories** item on the left-hand menu, and you'll find something similar to this:

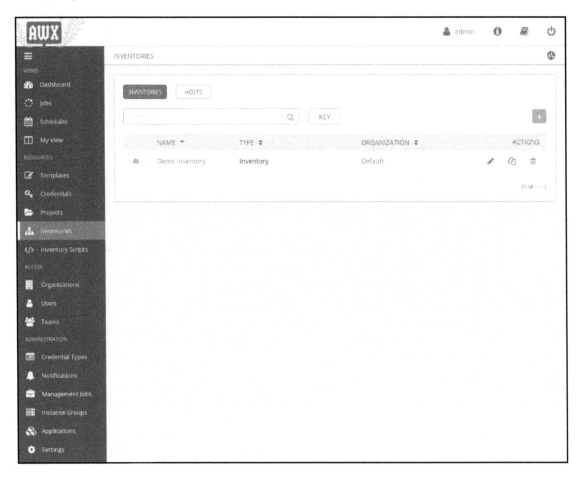

As for the projects, AWX comes with a demo inventory.

Looking from left to right, we can find the following columns:

- A cloud symbol – for inventory synchronization status
- The usual circle to show the status (okay or failed)
- The inventory **NAME**
- The inventory **TYPE**
- The **ORGANIZATION** that owns the inventory
- The edit symbol
- The duplicate symbol
- The delete symbol

As before, the green + button will allow you to create a new item. By clicking it, it will ask you if you want to create an inventory or a smart inventory.

We can select the **Inventories** option for now, and it will allow you to add a name and an organization (the only two mandatory options) as well as other non-mandatory options. As soon as you save, you'll be able to add hosts, groups, and permissions.

If you prefer not to specify the hosts, groups, variables, and so on by hand, there is a **Sources** tabs for you.

By clicking the + on the **Sources** tab, you'll be able to add a source from a list of available types or with a custom script.

The available sources types at the time of writing are as follows:

- **Sourced from a project**: Basically, it will import an Ansible core inventory file from a repository.
- **Amazon EC2**: It will use the AWS API to discover all the EC2 machines and their characteristics running in your environment.
- **Google Compute Engine (GCE)**: It will use the Google API to discover all GCE machines and their characteristics running in your environment.
- **Microsoft Azure Resource Manager**: It will use the Azure API to discover all machines and their characteristics running in your environment.

- **VMWare vCenter**: It will use the VMWare API to discover all machines and their characteristics managed by your vCenter.
- **Red Hat Satellite 6**: It will use the satellite API to discover all machines and their characteristics managed by your satellite
- **Red Hat CloudForms**: It will use CloudForms API to discover all machines and their characteristics managed by it.
- **OpenStack**: It will use the OpenStack API to discover all machines and their characteristics running on your OpenStack environment.
- **Red Hat Virtualization**: It will use the RHEV API to discover all machines and their characteristics running on it.
- **Ansible Tower**: It will use another Ansible Tower/AWX installation API to discover all machines and their characteristics managed by it.
- **Custom script**: It will use a script that you uploaded in the *Inventory Scripts* section.

We have seen now how to set up an AWX Inventory, which is going to be needed for the next part: setting up an AWX job template.

Understanding AWX job templates

In AWX, we have the concept of a job template, which is basically a wrapper around a playbook.

To manage the job templates, you have to go to the **Templates** section from the left menu, and you'll find something like the following:

Looking at the table containing the job templates, we will find the following things:

- The job template name
- The template type (AWX supports also workflow templates, which are templates for groups of job templates)
- The rocket button
- The duplicate button
- The delete button

By clicking the rocket button, we can execute it. Doing so will automatically bring you into a different view, which we will discover in the next section.

Using AWX jobs

AWX jobs are executions of AWX jobs templates, in the same way as Ansible runs are executions of Ansible playbooks.

When you launch a job, you'll see a window just like the following one:

This is the AWX version of the output of Ansible, when run on the command line.

After a few seconds, in the right-hand grey box a very familiar output will start to pop out, since it's exactly the same `stdout` of Ansible, just redirected there.

If later you click on **Jobs** on the left menu bar, you will find yourself on a different screen, listing all previously run jobs:

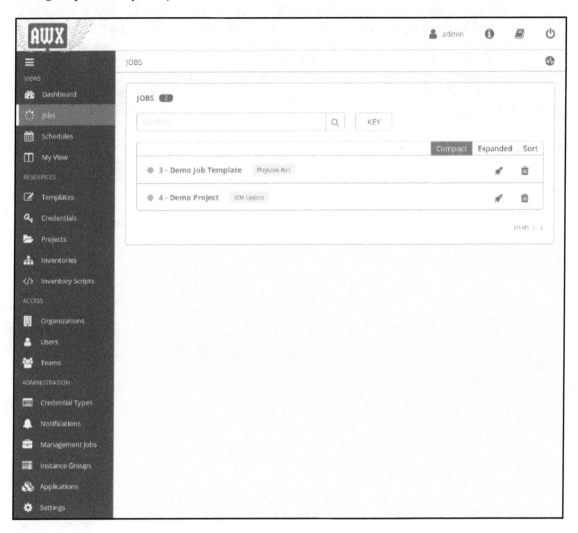

As you can notice, we have two jobs that have been executed, while we've only executed the **Demo Job Template**. This is because the **Demo Project** has been pulled before and due to the **Demo Job Template** execution. This allows the operator to be always comfortable to run a job, knowing that it will always be the latest version available in the SCM to be executed.

Summary

In this chapter, you have learned to set up AWX on Fedora and you've learned to use AWX projects, inventories, jobs templates, and jobs. As you can imagine by the number of options, flags, and items present in AWX, this is just the tip of the iceberg and does not intend to be a full explanation of it, since a dedicated book would be needed for that.

In the following chapter, we are going to discuss a little bit about AWX users, users permissions, and organizations.

12
Working with AWX Users, Permissions, and Organizations

While reading the previous chapter, you probably asked yourself questions about the security of AWX.

AWX is very powerful and, to be so powerful, it needs a lot of access to target machines, which means that it can become a potentially weak link in the security chain.

In this chapter, we are going to discuss a little bit about AWX users, permissions, and organizations; namely, we are going to cover the following topics:

- AWX users and permissions
- AWX organizations

Technical requirements

To fulfill this chapter, we will only need AWX, which we set up in the previous chapter.

AWX users and permissions

First of all, if you remember the first time you opened AWX, you will remember that you had to input a username and password.

As you can surely imagine, those were default credentials, but you can create all of the users that your organization needs.

To do so, you can go to the **Users** section in the left-hand side menu, as shown in the following screenshot:

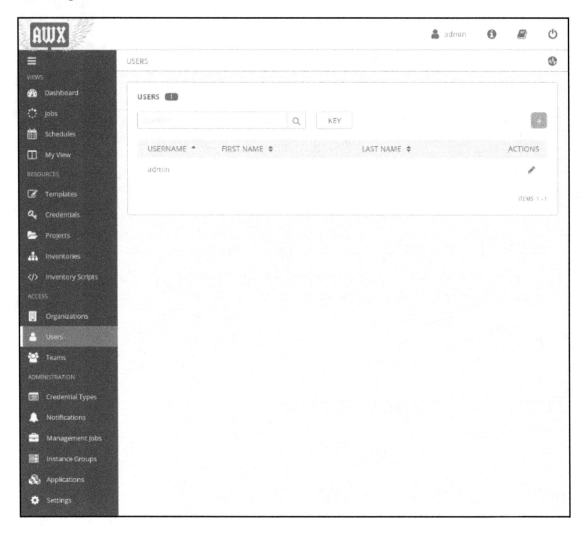

As you might expect, the **admin** user is present, and it is the only present user.

We can create other users by clicking the green button with the + symbol on it.

When we create a new user, the following fields are requested:

- **First name**: This is the user's first name.
- **Last name**: The user's last name.

- **Organization**: The organization that the user belongs to (we will speak more about this later in this chapter).
- **Email**: This is the user's email.
- **Username**: This is the user's username. This will be used for logins and it will pop up in the user interface.
- **Password**: This is the user's password.
- **Confirm password**: Re-type the password to ensure that no typos slipped in.
- **User type**: A user can be a normal user, a system auditor, or a system administrator. By default, normal users don't have access to anything, if not explicitly granted. System auditors can see everything in the whole system, but only in read-only mode. System administrators have full read-write access to the whole system.

After you have created a user as a normal user, you can go to **Templates**. If you go into the Edit mode for `Demo Job Template`, you'll notice a **Permissions** section, where you can see and set the users that are able to see and operate on this job template. You should see something like what's shown in the following screenshot:

By clicking on the green button with the + symbol, a modal will appear where you can select (or search) for the user that you want to enable, and you can choose the access level, as shown in the following screenshot:

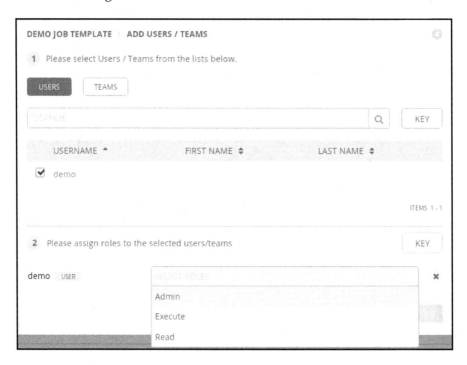

AWX allows you to choose between three different levels of access:

- **Admin**: This kind of user is able to see the job template and the past jobs that were created with it, execute the job template in the future, and edit the job template.
- **Execute**: This kind of user is able to see the job template and the past jobs that were created with it and execute the job template in the future, but not to edit the job template.
- **Read**: This kind of user is able to see the job template and the past jobs that were created with it, but is not to execute it nor change it.

In the same way job templates can be made visible, usable, and manageable by users, all of the other kinds of objects present in AWX can have permissions as well.

As you might imagine, if you start to have tens of jobs and tens of users, you'll spend a lot of time managing permissions. To help you with this, AWX provides the concept of teams.

Teams are manageable in the **Team** item of the left-hand side menu, and are basically just groups of users so that you can move from a **Discretionary Access Control** (**DAC**) approach to a **Role-Based Access Control** (**RBAC**) approach, which is much quicker to keep up to date with regarding organizational changes and needs.

By using users, teams, and permissions, you'll be able to decide who is able to do what at a very granular level.

AWX organizations

In more complex organizations, is often common that many people belonging to very different teams and business units share the same AWX installation.

In those cases, it can make sense to have different AWX organizations set up. This allows for easier permissions and the delegation of some permission management from the core system administrators team to organization administrations. Also, organizations allow for vertical permissions on the organization resources, such as inventory administrator (that is, the automatic administrator of all inventories owned by the organization) or project admin (that is, the automatic administrator of all projects owned by the organization), aside from the organization-wide roles (such as organization administrator and organization auditor).

If you are in a company that has multiple websites, you could decide to cluster all websites in the same AWX organization (if they are managed by the same people, for instance, the "web group"), or you could decide to split them into multiple AWX organizations, one for each website.

The advantages that are brought by these organizations are as follows:

- Easier permission management
- Ability of the team manager (that is, the "web group" manager or the single website manager) to on-board and off-board members as time goes by
- Easier and quicker audits, since only the permissions related to the specific organization will need to be vetted, instead of all permissions in Tower

Due to those advantages, I always suggest that you think about how to use an AWX organization in AWX.

Also, in my experience, I've always noticed that the more similar the AWX organization's structure is to the company structure, the better it is, since it's going to feel natural to all users. On the other hand, if you try to force an AWX organization's structure that is completely different from the company structure, this will feel alien, will slow down the adoption of AWX, and, in same cases, can even determine the failure of the platform.

Summary

In this book, we started from some very basic concepts of automation by comparing Ansible to the other common options such as manual procedures, bash scripting, Puppet and Chef are available. We then moved on and looked at how to write YAML files, since this is the format that's used by Ansible, and how to install Ansible. We then moved on and created our first Ansible-driven installation (a basic couple of HTTP servers backed by a database server). We then added features to leverage Ansible's features, such as variables, templates, and task delegation. Then, we moved on and saw how Ansible can help you in cloud environments such as AWS, Digital Ocean, and Azure. We then moved on to analyze how Ansible can be used to trigger notifications, as well as in various deployment scenarios. We closed with an overview of the official Ansible graphical interface: AWX/Ansible Tower.

Thanks to this content, you should now be able to automate all of the possible scenarios you'll encounter in your Ansible usage.

Other Books You May Enjoy

If you enjoyed this book, you may be interested in these other books by Packt:

Mastering Ansible - Third Edition
Jesse Keating, James Freeman

ISBN: 978-1-78995-154-7

- Gain an in-depth understanding of how Ansible works under the hood
- Fully automate Ansible playbook executions with encrypted data
- Access and manipulate variable data within playbooks
- Use blocks to perform failure recovery or cleanup
- Explore the Playbook debugger and the Ansible Console
- Troubleshoot unexpected behavior effectively
- Work with cloud infrastructure providers and container systems
- Develop custom modules, plugins, and dynamic inventory sources

Ansible 2 Cloud Automation Cookbook
Aditya Patawari, Vikas Aggarwal

ISBN: 978-1-78829-582-6

- Use Ansible Vault to protect secrets
- Understand how Ansible modules interact with cloud providers to manage resources
- Build cloud-based resources for your application
- Create resources beyond simple virtual machines
- Write tasks that can be reused to create resources multiple times
- Work with self-hosted clouds such as OpenStack and Docker
- Deploy a multi-tier application on various cloud providers

Leave a review - let other readers know what you think

Please share your thoughts on this book with others by leaving a review on the site that you bought it from. If you purchased the book from Amazon, please leave us an honest review on this book's Amazon page. This is vital so that other potential readers can see and use your unbiased opinion to make purchasing decisions, we can understand what our customers think about our products, and our authors can see your feedback on the title that they have worked with Packt to create. It will only take a few minutes of your time, but is valuable to other potential customers, our authors, and Packt. Thank you!

Index

B

bash modules
  using 154, 155, 156
blue/green deployment 209
Boto 65

C

canary deployment 208
CFEngine 8
check mode 165, 166, 167
cloud
  resources, provisioning in 104, 105
compiled software
  building, with RPM packaging 203, 204, 205, 206
Concurrent Versions System (CVS) 21
conditionals
  Boolean 80, 81
  variable is set, checking 81, 82
  working with 77, 79
Create, Remove, Update, or Delete (CRUD) 34
customized MOTD
  adding 47

D

data
  formatting, Jinja2 filters using 94
Database as a Service (DBaaS) 109
debugging commands 174
deployment strategies
  about 207
  blue/green deployment 209
  canary deployment 208
DigitalOcean 66
  deployment in 120, 121, 122
  machines, provisioning in 117
Discretionary Access Control (DAC) 241
DNS as a Service (DNSaaS) 108
Docker-py
  installing, in Fedora 226
Docker
  installing, in Fedora 226
Don't Repeat Yourself (DRY) principle 82
dry-run 165

dynamic inventory 57

E

Elastic Block Storage (EBS) 109
Elastic Compure Cloud (EC2) 107
emails
  sending, with Ansible 126, 127, 128
environment
  preparing 186, 188, 189
EPEL
  enabling 43
exceptions
  managing 175, 176, 177
execution strategies
  about 92
  free 93
  linear 93
  serial 93
exit_json
  working with 151, 152

F

fail_json
  working with 151, 152
failure
  triggering 177
Fedora
  Ansible, installing in 226
  Docker, installing in 226
  Docker-py, installing in 226
  user permission, giving to use Docker in Fedora 226
fevelopment 180
FirewallD
  ensuring 45, 46
free execution 93
Fully Qualified Domain Name (FQDN) 27

G

Git branch
  code 180
Git
  Ansible, using with 22, 23

X

Y

www.ingramcontent.com/pod-product-compliance
Lightning Source LLC
LaVergne TN
LVHW081520050326
832903LV00025B/1561